Contents

GrowerTalks®
on Pest Control

Edited by
Rick Blanchette

Ball Publishing
Batavia, Illinois, U.S.A.

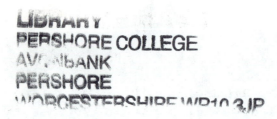
Ball Publishing
335 North River Street
Batavia, IL 60510, USA
www.ballpublishing.com

Cover design by Tamra Bell.

Library of Congress Cataloging-in-Publication Data

GrowerTalks on pest control / edited by Rick Blanchette.
    p. cm.
Includes index.
    ISBN 1-883052-26-2
    1. Pests—Control. I. Blanchette, Rick, 1966- II. GrowerTalks. III. Title.
    SB950 .G76 2001
    632'.9—dc21

00-012269

Printed in the United States of America
05 04 03 02 01 00 1 2 3 4 5 6

Contributing Authors

Anne Alvarez is professor, Department of Plant Pathology, University of Hawaii, Honolulu.

Larry W. Barnes is associate professor and extension plant pathologist, Texas A&M University, College Station, Texas.

Brent Bates is senior vice president, director of claims, Florists' Mutual Insurance Company, Edwardsville, Illinois.

Kurt Becker, greenhouse technical sales, The Dramm Corp., Manitowoc, Wisconsin.

Mike Benson is professor of plant pathology, North Carolina State University, Raleigh.

Chris Beytes is editor of *GrowerTalks* magazine, Batavia, Illinois.

Cassy Bright is varieties manager, Van Zanten North America, Oxnard, California.

Sherri Bruhn is editor of *Seed Trade News,* Batavia, Illinois.

Karol A. Burns is a graduate research assistant, Department of Entomology, Texas A&M University, College Station.

Leslie Campbell is senior research associate, University of California, Davis.

David Cappaert is technical advisor at Koppert Biological Systems, Romulus, Michigan.

Christine Casey is graduate research assistant and Ph.D. candidate, University of California, Davis.

A. R. Chase is president and plant pathologist, Chase Research Gardens Inc., Mt. Aukum, California.

Mike Cherim is director, The Green Spot Ltd., Nottingham, New Hampshire.

David L. Clement is director, Home and Garden Information Center, University of Maryland Cooperative Extension Service, Ellicott City.

Raymond A. Cloyd is assistant professor/extension specialist in ornamental entomology/IPM, Department of Natural Resources and Environmental Sciences, University of Illinois, Urbana.

Paul D. Curtis is senior extension associate and coordinator of the Wildlife Damage Management Program, Department of Natural Resources, Cornell University, Ithaca, New York.

Margery Daughtrey is senior extension associate, Long Island Horticultural Research Laboratory, Cornell University, Riverhead, New York.

Bob Decker is assistant vice president of loss control, Florists' Mutual Insurance Company, Edwardsville, Illinois.

Bastiaan M. Drees a professor and extension entomologist, Department of Entomology, Texas A&M University, College Station.

Ethel Dutky is plant pathologist, Department of Entomology, University of Maryland, College Park.

Wade H. Elmer is associate plant pathologist, Connecticut Agricultural Experiment Station, New Haven.

Joe E. Flaherty is assistant in bacteriology, University of Florida Gulf Coast Research and Education Center, Bradenton.

Hisae Fukui, University of California, Riverside.

Ryo Fukui, University of California, Riverside.

David M. Geiser is assistant professor of plant pathology and director of the Fusarium Research Center, Pennsylvania State University, University Park.

Hans S. Gerritsen, The Hortus Group, Salinas, California.

Stanton Gill is regional specialist, Central Maryland Research and Education Center, University of Maryland Cooperative Extension, Ellicott City, and adjunct professor, Montgomery College.

Lynn P. Griffith Jr. is president, A&L Southern Agricultural Laboratories, Pompano Beach, Florida.

Roberto Gonzalez is research and development manager, Van Zanten North America, Oxnard, California.

P. Allen Hammer is professor, Department of Horticulture and Landscape Architecture, Purdue University, West Lafayette, Indiana.

Brent K. Harbaugh is professor of floriculture, University of Florida Gulf Coast Research and Education Center, Bradenton.

Mary K. Hausbeck is assistant professor, Department of Botany & Plant Pathology, Michigan State University, East Lansing.

Will Healy is manager of technical services, Ball Seed Company, West Chicago, Illinois.

Kevin M. Heinz is associate professor, Department of Entomology, Texas A&M University, College Station.

Harry A. J. Hoitink is professor, Department of Plant Pathology, The Ohio State University, Wooster.

R. Kenneth Horst is professor, Department of Plant Pathology, Cornell University, Ithaca, New York.

Jae-soon Hwang is a Ph.D. graduate student, North Carolina State University, Raleigh.

Dr. Lee Jackson is research scientist, AgriPhi Inc., Logan, Utah.

Dr. Jeffrey B. Jones is professor of bacteriology, Department of Plant Pathology, University of Florida, Gainesville.

Mike Klopmeyer is pathology group manager, Ball FloraPlant, West Chicago, Illinois.

Dr. James F. Knauss, technical manager, Technical Service Group, The Scotts Co., Marysville, Ohio.

Edwin Lewis is assistant professor of turf and landscape entomology, Department of Entomology, Virginia Tech, Blackburn.

Jianbo Li is postdoctoral research scientist, Department of Entomology, Texas A&M University, College Station.

Richard Lindquist is professor, Department of Entomology, The Ohio State University/Ohio Agricultural Research and Development Center, Wooster, Ohio.

Pablo Bielza Lino, Departamento de Ingenierá Aplicada, Área de Producción Vegetal, Escuela Técnica Superior de Ingenieros Agró-nomos, Universidad de Murcia, Cartagena (Murcia), Spain.

Jim Locke is research plant pathologist, Floral and Nursery Plants Research Unit, U.S. National Arboretum, Beltsville, Maryland.

Joanne Lutz is president and IPM scouting specialist, Joanne's IPM Inc., Wheaton, Maryland.

Rosemary McElhaney is instructor, University of Hawaii's Kapiolani Community College, Honolulu.

Robert J. McGovern is associate professor of plant pathology, University of Florida, Gainesville.

Tunyalee Morisawa is a graduate student, University of California, Davis.

Brook C. Murphy is postgraduate researcher, Department of Entomology, University of California, Davis.

Joseph C. Neal is professor of weed science, Department of Horticultural Science, North Carolina State University, Raleigh.

Wendy O'Donovan is coordinator, clean stock/research & development, Oglevee Ltd., Connellsville, Pennsylvania, and a plant pathologist.

Ronald D. Oetting is professor of entomology, University of Georgia Experiment Station, Griffin, Georgia.

Denise L. Olson is a postdoctorate student, Entomology Department, University of Georgia Experiment Station, Griffin, Georgia.

Michael P. Parrella is professor and chairman, Department of Entomology, University of California, Davis.

Carol Puckett, Raleigh, North Carolina.

Karen K. Rane is plant disease diagnostician, Department of Botany and Plant Pathology, Purdue University, West Lafayette, Indiana.

Michael J. Raupp is professor, Department of Entomology, University of Maryland, College Park, Maryland.

Rondalyn M. Reeser is TPM/IPM scout-technician, University of Maryland Cooperative Extension Service, Ellicott City.

Karen Robb is farm advisor for floriculture and nursery crops, University of California Cooperative Extension Service, San Diego County.

Clifford S. Sadof is associate professor, Entomology Department, Purdue University, West Lafayette, Indiana.

John P. Sanderson is associate professor, Department of Entomology, Cornell University, Ithaca, New York.

Cheryl Smith is plant health extension specialist, University of New Hampshire Cooperative Extension, Durham.

John Speaker is an independent IPM scout and owner of Speaker's Garden, Silver Spring, Maryland.

Dr. Roger C. Styer is president, Styer's Horticultural Consulting Inc., Batavia, Illinois.

Brandi D. Thomas is sales assistant, *FloraCulture International,* Akersloot, the Netherlands.

Steve Thompson is staff research assistant, Department of Entomology, Texas A&M University, College Station.

Tom Thomson is product development vice president, Monterey Chemical Co., Fresno, California.

Diane Ullman is professor of Entomology, University of California, Davis.

Dave von Damm-Kattari is staff research associate, Department of Entomology, University of California, Davis.

Gina K. von Damm-Kattari is postgraduate researcher, Department of Entomology, University of California, Davis.

Leslie R. Wardlow is a horticultural pest control consultant from Kent, England.

Anna Whitfield is graduate research assistant and MS candidate, University of California, Davis.

Don C. Wilkerson is professor and extension horticulturist, Department of Horticultural Sciences, Texas A&M University, College Station.

Jean L. Williams-Woodward, University of Georgia, Athens.

Jim Willmott is extension specialist, Rutgers University Cooperative Extension Service, Clementon, New Jersey.

Introduction

Pest control is an issue that growers must face year after year. It never goes away. Each new season, each new crop brings its own set of concerns—thrips, aphids, fungal diseases, viruses, etc. Any one of these can make the difference between commercial success and financial disaster.

And the methods to fight these problems keep changing too. New chemicals are registered, while others are discontinued or banned, and you need to find chemicals to rotate so pests don't build up resistance to any one class. Biological controls, too, are changing, as more research is conducted on the best natural enemies and beneficial organisms to use.

We've included in *GrowerTalks on Pest Control* the best articles from *GrowerTalks* magazine on the subject of pest and disease control. These entries were written by leading academics doing the latest research and by top growers who have fought the same fights you fight—and have won. Learn from their victories and grow the best crops you can!

While we have tried to organize these chapters topically, note that many subjects overlap. An article in the chapter Specific Pests may deal with fungus gnats, but it includes information relating to biological control, IPM, or chemical control. An article in the Chemical Control chapter will discuss pests or diseases that might not be addressed in the general pest or disease chapter. Check the table of contents carefully to make sure you're not missing any articles that you need. The index in the back of the book will provide you with a complete listing of pests, diseases, and crops to aid you in your search.

Also, please note that these articles span a period of a few years. Chemical availability and labeling change constantly and vary regionally, so a registered product may not be currently in your arsenal, or an experimental chemical may be commercially available to you now. Plus, many of the research articles experiment to evaluate new products or find new uses for existing products. Please be careful when selecting and using any of the products referred to here. Check your local ordinances and follow the labeling. As with any new treatment, trial first before using on an entire crop.

Chapter 1

Integrated Pest Management

Using Active Scouting to Demystify IPM

Ronald D. Oetting

I was in a meeting a few weeks ago when professional entomologists discussed a definition of integrated pest management (IPM). There was a common thread: integrate different methods to control pests. However, there were mixed opinions on the primary emphasis of which method to use and the role of pesticides in IPM. The definition of IPM varies from using pure biological control to developing better tactics of using pesticides.

The mystery of how to define IPM will continue, but the message to growers is that it's time to start improving your pest control program by becoming more aware of what alternatives are available and how they might fit into your plant production.

Scouting Techniques

Scouting is the foundation for developing an IPM program. Growers need to develop some type of scouting program to obtain information as a basis to better manage pests. What is scouting? We'll use the term "scouting" to describe methods of checking a crop for pests and developing information for a decision on techniques to keep pest populations at an acceptable level.

There are both active and passive methods of looking for pests in the greenhouse. Active scouting involves the actual searching of the plants and surrounding environment for evidence that pests are present in the crop. Passive scouting includes using trapping, such as yellow sticky cards, to capture pests as they move within the crop.

Selecting a Scout

Start a scouting program by assigning the task to someone who has the responsibility of regularly checking the greenhouse for pests. This individual must know what pests to look for and how to look for them. It's his or her responsibility to collect the information used to determine if pests are present and if action is needed. Develop a written form for recording this information on pests' presence and maintain the forms as a permanent record of pest occurrence. All areas of the greenhouse should be checked and the information recorded on this form. This form should be simple and easy to use. It should record the date of scouting, the crop, the greenhouse, insects and diseases found, a map of the greenhouse, observations that need to be checked later, changes observed, and pesticide usage. The information will be used to make decisions on current activities and will be used in the future to observe trends, predict areas of pest occurrence and timing, and to develop data to be used in making better decisions.

Developing an Action Plan

What must be done before scouting? The first step is to develop a schedule to weekly scout each greenhouse and make it part of the work plan. Develop a plan to use as a standard method of searching for pests within a greenhouse. Divide the greenhouse into equal areas, and use a pattern of movement through the house so that you'll check plants in each area. Pests often enter the greenhouse at doors, bench ends, and near cooling pads. There's also a tendency for pests to build up in corners, areas of the greenhouse that have higher or lower temperatures or moisture, and other areas not frequented by workers, where they can increase without being detected. These are key areas that become more obvious with experience.

Modifying the Search

The scout should make it a standard practice to check the area in and around the greenhouse before plants are brought into the house. Look for weeds, holes in plastic, or other possible things that might contribute to insects or diseases being present. It's best to take care of these items before introducing plants to the area.

Start checking the plants by walking the aisles between the benches, checking plants in each area, using the techniques for the pests that occur on that crop. One or two key pests for each crop will dictate the selection of techniques. This will be the insect or disease most difficult to control or, for some reason, that threatens the production of a quality product. Concentrate on checking for key pests, and at the same time look for other pests that might be present. It's important that a minimal number of plants are checked and that this information is recorded for immediate or later use. This will provide a standard record that can be used in modifying and improving your scouting program. In addition to looking at the top of plants, periodically check the root ball for root growth and abnormalities.

Visually inspect the roots for browning or other deterioration that could indicate root rots.

Don't forget to check under benches for weeds. Any abnormality should be recorded, and if the cause is unknown, samples should be collected and taken to someone for determination of the problem's source. It's a good idea to develop a technique of marking areas where problems are found. This can be done by marking a map of the greenhouse on the scouting form or placing a flag or other marker on the bench. Replace yellow sticky cards with clean ones, and record the number of insects per card in the proper place on the form.

Choosing a Technique

Each crop requires different scouting techniques. Knowing the key pests that frequent a particular crop will be essential in determining what scouting methods are used. Different insect and mite pests have characteristics that are used in looking for their presence. Why? Because they're different in the way they move into a crop.

Different distribution characteristics of pests include:

Random

This group is often randomly distributed in the greenhouse very soon after entry and before being detected. It includes pests such as thrips that fly and move from plant to plant.

Spots

These pests aren't as mobile, and the population increases in size before becoming distributed throughout the greenhouse. They either can't or don't fly once inside the house. For example, mites don't fly and are brought into the house on plants or workers. They're located near the edge of the house where they entered by crawling into the house or by being carried in on air currents. Aphids have a winged form that can fly, but when an individual lands on a plant, the offspring change to a wingless form. They increase in population until a crowded level is reached before new winged forms are produced. These individuals often fly for a short distance and start new colonies, or they may move farther away from the source and start new colonies in other areas of the house.

Patches

Some pests can fly and, upon entering a greenhouse, will lay eggs on plants in one area of the greenhouse. This will result in a patchy infestation in one or more areas and include some of the moths that lay eggs concentrated on plants in one area.

Marginal

This includes many pests that move into a crop from the outer edge, near an opening to the greenhouse, or end of a bench or row.

Spotting the Big Three

Western flower thrips (WFT)

The western flower thrips is one of many thrips that can be found on ornamental crops. It's the most significant of the thrips because it's very difficult to control with insecticides and can vector virus diseases to ornamental crops. It's native to the western states and has only been a problem east of the Mississippi River for about fifteen years. One crop with thrips problems is pot-grown chrysanthemums.

The detectable characteristics of WFT are location on the plant and type of damage to look for. The location on the crop changes with the growth stages of the plant. On young plants, thrips feed on the new leaves, and immature thrips develop there.

The method to scout for thrips is to look for damage. The probability of seeing the thrips is nearly impossible without close inspection with a hand lens. The damage is expressed as light spots or distorted growth on the leaf and can usually be seen on the upper leaf surface. The spots are usually irregular in shape (not the pinpoint spots typical of mite feeding). The region is often sunken, and there may be black spots, fecal drops of the young thrips, in the area. In some of the feeding spots, you can detect young thrips, which are white to yellow and visible with a 10X hand lens. To scout, therefore, pick up a pot and turn it in your hands, looking at the upper surface of the leaves, especially the newest growth. Repeat looking at pots at random throughout each section of the greenhouse (probably ten to twenty pots), and record your findings on the scouting sheet.

As the plant grows and the leaves harden, thrips damage isn't as easy to detect. When buds first appear, look for brown spots on the new buds. As the buds enlarge, feeding will cause dark spots on the top of the bud and distort bud opening. When the flowering stage is reached, both damage and all stages of thrips can be detected. A good way to sample for thrips in flowers is to lightly blow into the flower. Carbon dioxide in your breath will cause the thrips to move around, increasing your ability to see them.

Another method of checking flowers is to tap flowers over a white tray and then check for thrips on the tray.

Thrips are general in distribution, so they'll probably be throughout the crop and not in a isolated area. Check for WTF with a different technique at different stages of plant growth. In addition to actively searching for thrips, yellow or blue sticky cards can also be used to detect thrips' presence and provide an indicator of numbers present in the crop.

Silverleaf whitefly

The silverleaf whitefly has been a pest on greenhouse crops for over ten years and is the most common pest on poinsettias in the fall. In scouting for whiteflies, turn leaves to look at the underside for immature and adult whiteflies. They don't cause damage that can be easily detected, so you must look for the individuals on the plant. The adults look like small white flies (thus the common name) and are easy to detect, but the immatures are more difficult to see. They are scalelike and cream-to-clear colored and are sessile on the leaf. Because they don't move and are tight against the leaf surface, they blend in and aren't easy to detect until the numbers are high and the leaf is speckled with individuals.

The lowest leaf of the plant, which is flat against the top of the pot and potting media, is where the initial population of whiteflies is located. These leaves are hard to reach with sprays and should always be checked while sampling. In addition, leaves higher on the plant should be checked, but the most critical leaves for early detection are the lowest leaves on the plant. Low-level populations of whiteflies are often found in patches, and spot sprays can be used at this time. The population will spread over the crop when new adults emerge. It's common to have certain areas of the greenhouse where populations are always higher. The yellow sticky cards are good tools to detect whitefly presence, but once the population is established, they're of less value.

Two-spotted spider mites

Two-spotted spider mites have been major pests of greenhouse crops for many years. They're the most common pests found on ornamentals because almost everything grown in the greenhouse can be a host. They're found on the undersides of leaves, but because of their small size, it's hard to detect them.

To scout for spider mites, look for any abnormality on leaves, then check the underside of the leaf with a hand lens for mite presence. The most noticeable damage is a small pinpoint-sized spot observed on the upper surface. Usually there'll be a patch of these small spots on one area of the leaf. The mites are found on the underside and can be detected with a 10x hand lens. This damage will be in an isolated spot of the greenhouse, usually where it's warm and there's worker activity near doors. It's also common to find them in corners of the greenhouse, as these are areas that might not receive good coverage when treating the area with pesticides. Because mites first occur in isolated spots, early detection will allow growers to use spot sprays in the area where they first appear. Sticky cards are of no value in scouting for mites because they don't fly.

The development of a scouting program is important in good plant health management, independent of what type of control program is used. If you don't have scouting as part of your overall production strategy, consider it. If you do, it's probably time to review the progress and update the program.

January 1999.

Scouting Basics

Karol A. Burns, Kevin M. Heinz, and Bastiaan M. Drees

Pest outbreaks in greenhouses can result from the explosive growth of an existing population, by migration of a new population into the greenhouse from surrounding areas, or via new plant material or potting soil. Insects may also be introduced by "hitching a ride" on personnel moving between greenhouses.

But not all insects are harmful. Some are beneficial because they consume or lay eggs in the bodies of pest insects, killing them. If you maintain beneficial insect populations already present in the greenhouse, you can reduce pesticide use.

Carefully monitoring pests and beneficial insects is the backbone of IPM—integrated pest management. Here are some of the monitoring techniques used in greenhouses today. We'll also take a look at research we're doing at Texas A&M University to develop a simplified monitoring system that's statistically acceptable, saves time, and is easy to do.

Why Monitor?

There are many reasons to monitor plants for insect and mite pests:

To detect initial pest presence. Regular monitoring can detect when pests first appear in your greenhouse, giving you time to take action. For example, certain insects are responsible for transmitting plant viruses, such as thrips carrying impatiens necrotic spot virus (INSV).

To track pest growth stages. If you know when a pest population has reached a life stage susceptible to chemical applications, parasitoid/predator releases, beneficial pathogens, or insect growth regulator applications, you can make fewer but more effective control applications.

To determine when to take action. Knowing what insect species are present and their location and density can reduce the number of pesticide applications you need to make. This helps reduce the chance of pests developing resistance and minimizes the risk of environmental contamination.

To track the number of pests and natural enemies. If you know how many pests and natural enemies are in your greenhouse, you can more accurately determine what and when to apply for control and if previous control measures have been effective.

Know Your Foes—and Friends

The first step in any monitoring program is accurate pest and beneficial insect identification. Overly generalized identifications can lead to control problems. For example, developing stages of fungus gnats and shore flies occur in different habitats and require different control approaches and treatments.

Second, knowledge of the pest's life cycle is essential for properly timing your chemical treatment or beneficial insect release, greatly improving your treatment effectiveness.

Insect identification and biology are topics too large for inclusion in this article, but many good reference books are available, most with excellent color photographs. In the table, we have summarized the damaging stage(s) for six of the most common greenhouse pests, recommended sampling techniques for each stage, and the greenhouse crops typically preferred by each species.

Putting Together Your Program

Initiating a monitoring program requires you to make several decisions:

Step 1: Who should scout?

Operations with successful scouting programs often use employees devoted to this purpose, or they may hire private IPM consultants. An in-house employee will be more familiar with the crop and growing system but may lack sufficient training. In-house employees also tend to be pulled away from

Suggested Sampling Methods for Greenhouse Pests

Common greenhouse insect pests	Affected crops*	Harmful stage(s)	Sampling method
Whiteflies *Trialeurodes vaporariorum, Bemisia argentifolia*	Tomato, sweet pepper, primula, poinsettia, hibiscus, gerbera, fuchsia, begonia, dahlia, coleus	Nymphs, adults	Timed counts, sticky traps
Aphids *Myzus persicae, Aphis gossypi*	African violet, begonia, carnation, nasturtium, rose, tulip, mums	Nymphs, adults	Timed counts
Spider mites *Tetranychus urticae*	Azalea, hydrangea, impatiens, freesia, fuchsia, salvia, marigold, dahlia	Nymphs, adults	Timed counts (hand lens required)
Thrips *Frankliniella occidentalis*	Tomato, sweet pepper, cucumber, mum, rose, impatiens, ivy geranium, petunia, gloxinia, orchid, dahlia, primula, gerbera, fuchsia, African violet	Larvae, adults	Beat sheet
Leafminers *Liriomyza trifolii, L. sativae, L. huidobrensis*	Wide host range (mums), wide host range (not mums), primula, dianthus	Larvae	Timed counts ("live" mines)
Fungus gnats *Bradysia spp.*	Fungi, decaying organic matter, plant roots	Larvae	Potato disks

*Not an all-inclusive list

regular scouting duties during busy seasons or personnel shortages. The advantages of a professional scout are that regular scouting is their only responsibility and they may be alert to new or developing problems at other operations in your area. One person should perform all sampling for consistent results. If two or more people sample, they should check each other in a preliminary sampling exercise to make certain their methods and results are similar.

Step 2: Identify your monitoring units

This may consist of the entire greenhouse, specific benches or bays, crops of the same age or cultivar, square feet, or some other logical layout. Hot-spot areas near doors, intake vents, and areas housing sensitive plant cultivars should be separate monitoring units. A sketch of the monitored area is helpful when selecting the coordinates of plants for sampling and for detecting any hot spots in the main crop.

Step 3: Choose a sample unit

The sample unit consists of plant parts where insect populations are typically found, such as leaves, terminals, stems, or roots. A common sample unit is employed for plants similar in structure—for instance, poinsettias and foliage plants. Leaf sample units are often used for these relatively large-leafed plants. Terminals are the more appropriate sample unit on small-leafed plants such as Mexican heather and plants with tight flower buds such as mums and carnations.

The sample unit you select should also be consistent with the detectable stages and feeding habits of the pest. For example, inspect middle and lower leaves for whitefly nymphs and pupae instead of young leaves, where difficult-to-see eggs and very young nymphs are typically found. Similarly, scale insects such as soft scales, armored scales, and mealybugs can be found on nearly all plant parts, including the stem.

Step 4: Choose a sample technique

The sampling technique is the method by which information is collected from a single sampling unit. There are many techniques to choose from, but it's critical that a standardized procedure be used throughout the crop cycle so that results can be compared. Commonly employed sampling techniques are:

Counting insects on plant parts

Typically, the sample unit is removed from the plant and the insects counted, either using the naked eye or a hand lens. This is an effective method for sampling aphids, spider mites, and other pests that don't fly or drop from the plant when disturbed. The number of samples removed ranges from five to twenty-five, depending on the plant species, size, and age. Large plants with many leaves can afford to have more of them removed than young or small plants with few leaves. Plant cultivars with known sensitivity to certain insect pests and older plants that have been in the greenhouse for some time should be sampled more aggressively. Because plant parts are typically removed, this method isn't the best choice for poinsettias or ornamental foliage crops that are highly valued for their aesthetic quality. For these, try timed counts.

Timed counts

Here, the scout counts the insects during a one- to two-minute visual inspection of an entire plant, including stems and the underside of leaves. Some experts recommend inspecting each plant for thirty seconds. A timer is required since it's difficult to count and keep track of elapsed time simultaneously, and timed counts are impractical when pest populations are high because you can't count the insects fast enough. The advantage of timed counts versus sampling plant parts is that plants aren't damaged. Use this technique for relatively stationary insects such as aphids, spider mites, leafminer larvae, and juvenile whiteflies.

Beat samples

With this technique, you beat or agitate the plant terminals or foliage a standard number of times (usually three to five) above a white surface such as a tray, piece of cardboard, or sheet of heavy paper. Count fallen arthropods immediately. This method is useful for thrips and other small insects that drop from the plant when disturbed, especially on plants with numerous tiny leaves, such as Mexican heather, or with tight flower buds, such as mums.

Traps

These devices contain synthetic or natural attractants called pheromones or use UV light or attractive colors to physically trap insects. Colored sticky traps will capture winged adults of thrips, fungus gnats, and whiteflies. UV lights and pheromone lures are sometimes used to detect the presence of moths. For best results, hang traps vertically just above the crop canopy, about one per 1,000 sq. ft. (use at least three traps in greenhouses smaller than 3,000 sq. ft.). Change traps weekly. Interestingly, capture rates of adult fungus gnats have been improved by placing sticky traps horizontally between the potting media and lowest leaves.

Attractant traps aren't a control tactic, nor can they help predict pest density. They capture only the adult stage (often only the male) even when no damaging immature stages occur on the crop. Use them for early detection of adult pests or of sudden population increases between scouting intervals.

Sentinel or indicator plants

Although not a scouting technique for determining pest population density, sentinel plants are useful as an early-warning device. The foliage of petunia varieties 'Summer Madness', 'Super Magic Coral', and 'Red Cloud' will show INSV symptoms and feeding damage from thrips before they occur in your main production areas. To enhance the sentinel plant's attractiveness to thrips, attach a yellow or blue nonsticky card to a stake and place it in the pot. Keep petunia flowers removed to prevent the pests from feeding on blooms.

You can also set aside or flag any infested plants found during regular inspections to determine the effectiveness of your control strategies during subsequent scouting intervals.

Potato disk method

This technique is occasionally used to monitor for fungus gnat larvae in the potting soil of container plants. Research at the University of Georgia has proven that adult counts from sticky traps are unreliable for estimating the true population density, compared with larval counts using potato disks.

To use them, put slices of raw potato on the soil surface for three to seven days. Larvae make their way to the soil surface to feed on the potato disk and can then be easily counted on the bottom of the disk. Potato discs are cumbersome and pretty fragrant after sitting in your greenhouse for a week; however, no other reliable alternative has been developed.

Step 5: Determine sampling pattern

Stratified random sampling is the technique of randomly sampling various units, either according to plant layers or sampling greenhouse spaces such as bays, benches, or square feet.

One way you can create a stratified random sampling procedure is to divide your monitoring unit into a grid. Next, you can use a random number generator (a feature available on many handheld calculators) to select the coordinates of individual sample units.

If you select plants only from the edges of benches located near walkways where personnel pass, you may overestimate the number of pests since movement into the greenhouse or sampling area occurs from the edges. You can avoid this by excluding perimeter plants as part of the monitoring unit.

Step 6: Determine sample timing

Time of day can affect the accuracy of population estimates. While some species are most active before noon, other insect species are less active in the cooler morning hours, affecting your counts. Also, flower odors from blooming crops are highly attractive to thrips and will draw them away from traps and nearby non-blooming plants.

Sample at least weekly. Ideally, scouting should occur frequently enough that an insect species doesn't pass through a developmental stage without being sampled. This interval will vary between species with different development cycles and can be further modified by the temperature in your area since insects develop faster in warmer climates.

Step 7: Determine optimal sample number

Generally, the number of samples you take from each monitoring unit at each sampling interval is held constant over the entire growing period.

Because pests aren't uniformly distributed throughout a crop, taking more samples results in more accurate population estimates. Time and equipment constraints may limit the number of samples, but the main point is consistency. The New York State Poinsettia IPM Program suggests inspecting twenty plants out of every two thousand for whitefly nymphs. At the University of Florida, scouts are trained to examine twenty plants per one hundred square feet of greenhouse space.

Step 8: Recording and using your data

Record your scouting results for each monitoring unit of the greenhouse. If possible, use a computer file to permanently store your data. (If you'd like a copy of one we've developed, please see the contact information at the end of the article.) This will give you easy access to previous years' data for comparison purposes. Records help you spot problems early enough to isolate or destroy pest-prone plants before the problem spreads. In future years, your records will help you know when to expect potential problems in a particular crop, as well as the insect development rates in your area. Comparing insect counts both before and after chemical applications or natural enemy releases helps determine if you achieved control. Spray records can be used to determine the pesticide, application method, and timing that resulted in the best control and least impact on beneficial insects. By comparing insect counts and crop quality, you can develop treatment thresholds based on your damage tolerance.

You can also calculate scouting labor and pest control costs and compare them to non-scouted areas. The cost savings achieved may help justify the hiring of additional scouts. Individual results will vary, but savings ranging from 24 to 39% have been documented for potted and cut flower crops even when the additional costs of scouting were included. In addition, pesticide spray volumes were reduced by a similar amount.

Soon to Come: A Simplified Routine

Despite the effectiveness of traditional monitoring programs, they typically address only one insect species per crop and often require prohibitively large sample sizes to be accurate. Implementing these programs quickly becomes confusing when multiple plant and insect species are present in the same greenhouse. In field or row crops, economic thresholds are well established so growers know when suppressive action is warranted. Few thresholds for greenhouse crops have been established.

This leaves you responsible for determining when to treat a crop based on individual pest or damage tolerance.

Our research at Texas A&M University is focused on simplifying the scouting routine in greenhouses by looking for commonalties among the distribution of six greenhouse pests that infest multiple plant species while still maintaining statistical accuracy. This new protocol reduces scouting time compared to existing methods. Here's how:

- It uses whole plants as the sample unit, thereby making the new protocol more universal among plant species similar in structure.
- It combines timed counts with a presence/absence sampling technique instead of counting individual insects per plant.
- It determines the optimal sample number for a range of statistical accuracy levels from which you can choose.
- It determines the minimum time interval for inspecting each plant, whether the objective is to quantify the entire number of pests present or for inspecting plants using the presence/absence technique.
- It uses portable handheld computers for data recording instead of paper forms, allowing results to be easily interpreted and stored on a desktop computer.
- It converts the percentage of infested plants to the actual number of insects per plant using a simple graph we've developed.

Field tests will tell us if our new scouting protocol produces the significant time savings, statistical accuracy, and ease of use we believe are possible.

The Computerized Scout

What do scouting and your local supermarket have in common? We believe computers and handheld scanners will play an increasing role in scouting programs of the future. Simple data recorders, already on the market, will eventually evolve into sophisticated scanners. When waived over a plant, they'll be able to recognize individual insect species and count them much like the scanner at the supermarket. Don't believe us? A scanner able to recognize whiteflies has already been developed by a company in Israel. But don't put off monitoring until then—start your own program today.

For more information, contact Texas A&M University, Department of Entomology, Fax: 409-845-7977.

October 1999.

TPM Part I: Forward-Thinking Pest Control

Stanton Gill

Total plant management—combining the framework of integrated pest management with a strong monitoring program—is today's smart solution for efficient and effective pest control

In the last two decades growers have been bombarded with major pests: western flower thrips—and the impatiens necrotic spot virus it can spread—silverleaf whitefly, the serpentine leafminer, *Botrytis,* and the root rot complex of diseases.

Their control has been made more difficult by pesticide overuse and the resulting resistance to one or more pesticides. Some growers have built a pesticide strategy that's like a house of cards—based on multiple, preventative applications of pesticides. This worked for a while, but now these strategies are collapsing as more and more pest controls fail.

The chemical companies are registering new pesticides, but growers must learn that relying solely on a chemical approach isn't only unwise, it's also expensive.

What's the solution for the forward-thinking greenhouse manager? Adopting a total plant management approach that uses the framework of IPM with a strong monitoring program is your best bet. In part one of our two-part series, we'll look at designing a TPM/IPM program that uses biopesticides and the proper way to scout.

Designing a TPM/IPM Program

The first step is to examine your crop and determine its key pests. The term "pest" includes insects, mites, diseases, and even cultural problems unique to each crop. For example, in impatiens the key pests are tospovirus (impatiens necrotic spot virus), *Pythium* and *Rhizoctonia* root rots, *Botrytis* blights, western flower thrips, excessive growth stretching early in the crop cycle, and excessive fertilizer rates. In the poinsettia TPM/IPM programs, the key pests are silverleaf whiteflies, greenhouse whiteflies, Lewis mites, dark-winged fungus gnats, *Pythium* and *Rhizoctonia* root rots, *Botrytis* blights, excessive plant stretching, and excessive or improper fertility.

Once the key pests are identified, you next put together a monitoring schedule and design data sheets to record monitoring data. Then you determine what other tools are needed for early detection and monitoring of the key pests.

An IPM program involves regular monitoring (in most cases at least weekly) and identifying problem pests and beneficial organisms present. With IPM, you identify pest problems early, isolate or destroy pest-prone plants before the pest problem spreads, develop treatment thresholds, use beneficial organisms where practical, and use spot treating with pesticides such as insect growth regulators and biorational products that have the least impact on beneficial organisms.

Market demand has made an unrealistic goal of eliminating all pests from the greenhouse. The reality is that retail evaluation studies have shown that few plants sold at retail are completely pest free. Pests are often present but at levels so low that most customers don't notice them. In most traditional chemical control programs, elimination of all pests is rare. The goal of most greenhouse pest management programs should be to maintain pests below an economically damaging level and at a level that your customer or state agriculture inspector will accept.

Combining Chemicals and Beneficials

Many new chemicals receiving registration for use in greenhouses have minimal impact on beneficial organisms, a very positive development for the greenhouse manager who uses beneficial organisms. When chemicals must be used to control plant-feeding pests, there are several ways to reduce potential adverse effects on beneficials. First, use only on plants or portions of plants requiring treatment. Second, apply materials at the time when they will be most effective against the pest. For example, do not spray a pest when the majority of the population is in the egg or pupa stage. Many pests have specific times in their life cycle when they are relatively immune to control by pesticides—do not apply pesticides during these times. Also, select pesticides that are least disruptive to the beneficial organisms you may be releasing in the greenhouse.

There are several biological pesticides (also called biopesticides) available to the nursery and greenhouse industry. The advantages of biopesticides over conventional chemicals are their selectivity to a targeted pest, lower toxicity to beneficial insects and greenhouse workers, and shorter restricted-entry intervals (REI) compared to conventional chemicals.

Monitoring Your Crop

Scouting

Your crop must be monitored on a routine, year-round basis. Successful scouting routines often use employees devoted just to that task or private

IPM consultants. The routine works best if it's coordinated through some-one designated to do the job. However, one major advantage to hiring an independent scout is having your crop examined weekly from a completely objective point of view. The scout provides you with a weekly report that can be used to make timely decisions concerning the crop's health.

Thorough scouting is best done by dividing your greenhouse into units of a size small enough to be inspected in a reasonable time—usually 2,000 to 4,000 sq. ft. It may be an entire 30-by-100-ft. greenhouse, or one bay of a gutter-connected greenhouse. Each unit is scouted weekly, with separate records kept for each.

Your scouting routine should include counts from several sticky cards and random foliage inspections to monitor non-flying pests. A few plants that have living pests can be flagged and used as sentinel plants, to be reinspected after sprays are applied or natural enemies released to determine if control was achieved. Finally, during each scouting session, assess the general health of the crop and note any other problems.

Sticky card monitoring

Use sticky traps to monitor insect populations on a weekly basis. Record the number of each insect pest captured on sticky cards to detect whether an insect population is building or declining. Also use sticky cards to detect if certain insects are migrating into your greenhouse. For general monitoring of most flying insect pests, place yellow sticky cards vertically on stakes, keeping the bottom of the card an inch or so above the top of the canopy. A rule of thumb is one or two cards per 1,000 sq. ft.

Foliage inspection

To monitor many insects, particularly non-flying ones such as spider mites, many aphids, caterpillars, weevils, scale and mealybugs, you must inspect the plants themselves. You'll find many pests on the underside of leaves. They may go unnoticed unless the leaves are turned over. Some pests may produce damage symptoms or other evidence of their presence such as honeydew, cast skins, fecal droppings, and ants feeding on the honeydew. The number of plants to inspect depends on the crop, the stage of crop, the pests that may be present, labor costs, and the experience of the scout, among other things.

In general, sample as many plants as is affordable, selecting randomly from throughout the growing area. Random plant selection is the best way to come across hot spots before they become obvious. Of course, if you know that certain crops, cultivars, or areas of the greenhouse are prone to pests, be sure to inspect a few plants from these areas. For general foliar scouting, select different plants each time the crop is inspected. Walk through the greenhouse and note any areas of the crops that are discolored, are of different height or shape, drooping, or have other subtle differences.

To inspect a plant, look at both leaves and stems, and if it's wilted, inspect the roots. Inspect a few leaves from the bottom, middle, and top of the canopy. Look for damage symptoms on the upper surface of leaves, but also turn some leaves over to check for pests.

It's important to record the monitoring results each week. A data sheet can be designed to keep track of weekly pest levels for each scouted area of the greenhouse. Include information on crop cultivar, crop stage, sticky card counts for each pest species, and counts from foliar inspections for each pest.

To determine the cost of your pest management program, record the time required for scouting, pesticide applications, and other pest management tactics. These records can then be used to determine if pest levels are declining. They can also be compared with spray records to determine the application method, pesticide, and time of application that gain the best control.

Available Biopesticides*

Product and active ingredient	Pest controlled	Comments
Avid (Abamectin) .15 E.C. formulation produced by soil microorganisms (*Streptomyces avermitilis*)	Leafminers, spider mites	Not considered disruptive to natural predators or beneficial insects. Takes three to four days to see maximum effectiveness.
Azatin E.C. (Azadirachtin)	Whiteflies, thrips, leafminers, leafhoppers, mealybugs, caterpillars, aphids	Mainly acts as insect growth regulator. May act as feeding and oviposition deterrent. May be used on herbs and vegetables.
Citation (cyromazine)	Dipterous leafminer larvae, fungus gnats and shore fly larvae in greenhouse crops	Insect growth regulator. Available in water-soluble packets.
Enstar II (kinoprene)	Aphids, whiteflies, scales, mealybugs, fungus gnats	Insect growth regulator.
Gnatrol (*Bacillus thuringiensis* var. *israelensis*)	Fungus gnat larvae	Applied to soil. Compatible with beneficial nematodes and hypoaspis mites for fungus gnat larvae control.
Mycostop	Controls seed rot, root and stem rot, and wilt caused by fusarium, alternaria, phomopsis; suppresses *Botrytis* infection	For use on container grown ornamentals and vegetables. Don't tank mix.

(continued)

Product and active ingredient	Pest controlled	Comments
Hot pepper wax	Acts as repellent against aphids, spider mites, thrips, leafminers, whiteflies, leafhoppers, scales	Contains hot pepper extract and paraffin wax concentrate with herbal extracts.
M-Pede	Aphids, mealybugs, leafhoppers, psyllids, scales, thrips, whiteflies, spider mites	Must make contact with insect—no residual effect. May be used on herbs and vegetable plants. Repeated applications to same plants can cause phytotoxicity.
BotaniGard (Mycotech Company) and Naturalis-O (Troy Chemical Company) (*Beauveria bassiana*)	Aphids, thrips, mites, whiteflies, leaf-feeding caterpillars, leafhoppers, psyllids	Needs to be applied as a fine mist and directly hit pests. Humidity levels must be above 35%.
Precision (Ciba company) (Fenoxycarb)	Whiteflies, aphids, thrips	Insect growth regulator. Twelve-hour REI.
Root Shield (*Trichoderma harzianum*) (BioWorks, Inc.)	Beneficial fungus that is antagonistic to *Pythium, Fusarium, Rhizoctonia*	Labeled for greenhouse crops and field crops.
Spod-X-L-C (*Spodoptera exigua*)	Beet armyworm	A nuclear polyhedrosis virus (NPV). Early instar larvae must digest product. Temperature above 90F inactivates the product.
SoilGard (*Gliocladium virens*, G1-21)	Beneficial fungus that's antagonistic to *Pythium* and *Rhizoctonia*, aiding in control of damping off and root rot pathogens	Suppression of root rot disease on vinca, geraniums and zinnia.
Scanmask and X-gnat (*Steinernema feltiae*)	Fungus gnat larvae	Will search for fungus gnats in stems and roots. Apply at temperatures between 50 to 90°F and apply as a soil drench.
SunSpray UltraFine oil (horticultural oil)	Controls aphids, whiteflies, mites, thrips, and labeled for powdery mildew control in greenhouses	Avoid applications on cloudy and humid days. Do not apply in sprayer used to apply fungicides. Fungicides and oil react, causing phytotoxic burn on some plants.
Triact 90 E.C. (A clarified hydrophobic extract of neem oil)	Black spot on roses, rust, powdery mildew, downy mildew, whiteflies, leafminers, thrips, caterpillars, mites, scales	For insects, spray at seven- to ten-day intervals. Don't spray on open impatiens flowers.

*This information is given with the understanding that no discrimination is intended and that no endorsement by the Maryland Cooperative Extension Service is implied.

It's critical to have a routine, organized plan for scouting results to go to the person responsible for making pest management decisions on a timely basis. The plan should include a follow-up mechanism to be sure that the control tactic, if needed, was applied and that the pest was controlled. Consult the above chart for a list of some of the available biopesticides.

February 1997.

TPM Part II: Forward-Thinking Pest Control

Stanton Gill

The days of relying solely on pesticides are gone, and resourceful growers are employing diverse pest management programs for repeated greenhouse success. Total plant management, which uses the framework of integrated pest management along with a strong monitoring program, is the forward-thinking grower's solution. Last month we looked at designing a TPM program using biopesticides and proper scouting techniques. In the final section of our two-part series, we'll tell you how to complete your TPM program by establishing thresholds and choosing the right biological controls.

Establish Realistic Thresholds

It's difficult to determine a threshold that's uniform for all growers. They produce many different crops, often simultaneously in various combinations in the same greenhouse. Also, outdoor climate differences make pest migration into the greenhouse a problem in some regions but not others. Finally, greenhouse structures vary tremendously.

You have to develop your own pest threshold, which may change as the crop grows. You can use your scouting records to determine your own thresholds. Note the number of pests you recorded on sticky cards or during foliar inspections at important times in each crop and whether the amount of damage was significant. Note whether you applied a control tactic and if the finished crop was acceptable to you and your customers.

Sanitation

Denying insects and mites their food sources is the best reason to practice sanitation. Sanitation means keeping your greenhouse weed-free and eliminating excess piles of growing media, algae, and plant debris. All of these provide insects and diseases with food and a place to live. Also, keep the area outside the greenhouse free of weeds and garden plants that can be sources of pests. Put compost or cull piles well away from greenhouse doors or vents.

Before starting a new crop, inspect the greenhouse for all existing pest problems. Don't forget stock plants, leftover hanging baskets, and weeds. Several viruses can be present in weeds and stock plants, even though these plants may not show symptoms. Thrips lay eggs on these infested plants, and larvae pick up the virus. After larvae pupate, virus-infected adults carry the pathogen to new plants. There's no cure for these viruses. Either eliminate these sources of pests or get the pest problem under control before you start the new crop.

Insect Screening

Microscreening your greenhouses to exclude pests can be an important part of your TPM/IPM plan, helping to reduce pesticide applications and all associated costs and reentry restrictions. Screening can help to fight insect resistance to pesticides by reducing the number of applications. Plus, a significant proportion of infestations come from insects moving into greenhouses from surrounding areas. Screening is about the only way to keep them out of your greenhouse.

The level of exclusion you need depends, in part, on what you grow. Propagators need to be especially stringent in their exclusion efforts to ensure that they don't ship infested plants. Propagation for your own use requires close to 100% exclusion of pests. This means restricting greenhouse access to key personnel who take careful measures to ensure insects aren't carried in on their clothing or plant material.

If you use microscreening on finished crops, you don't have to maintain as high a level of caution as for plant propagation, although it's very important that workers keep doors closed and don't move infected plant material into the house.

Many growers have achieved significant reductions in pest populations and reduced pesticide applications without 100% insect exclusion. Screening can also be effectively combined with biological controls. The University of Maryland Cooperative Extension has published fact sheet #185 on considerations for installing microscreening for commercial greenhouse operations. For copies, write: CMREC, University of Maryland, 11975 Homewood Road, Ellicott City, Maryland 21042. (*Editor's note: the National Greenhouse Manufacturer's Association has published "Greenhouse Insect Screening 1996." To order, Fax: (303) 798-1315.*)

Using Biological Controls

When released early in a crop cycle to keep pest populations from building up, biological control agents—also known as beneficial insects or beneficials—are an integral part of your TPM/IPM program. Commercially available beneficials control several major pests in greenhouses.

You can use predatory beneficials such as green lacewing (*Chrysoperla carnea*) that feed on a wide range of insects such as whitefly, aphid, mealybug and several species of caterpillar. Parasites such as *Aphidius colemani*, which parasitizes aphids, tend to target certain insect species. Several entomopathogenic fungi and nematodes are also available (see chart).

Biological control is best used with a strong weekly monitoring program. If pesticide applications are necessary, select pesticides that are compatible with beneficial organisms.

Many state university cooperative extension services have ongoing greenhouse TPM/IPM programs. If your state cooperative extension service doesn't have an IPM program, ask for advice on how to contact a local independent IPM scout.

Insect, Mite, and Disease Control			
	Best method to monitor	Identification features	Potential biological control
Insects			
Aphid (general)	Using sticky cards indicates aphid infestation or migration into greenhouses in spring, summer, and fall. Inspect plant foliage on a weekly basis. The presence of cast skins or honeydew is a good indicator.	Size, color, location on plant, and crop preference are different for each species. Most aphids are .04 to .16 inches in size, pear shaped, soft bodied, and have cornicles (tailpipes) present on the rear of the abdomen. Legs and antennae are typically long and slender. Winged forms are found on cards; wingless forms are found on plants.	Green lacewing (*Chrysoperla carnea* and *Chrysoperla rufilabris*), aphid midge (*Aphidoletes aphidomyza*), parasitic wasp (*Aphidius colemani*), parasitic fungus *Beauveria bassiana* (Naturalis-O and BotaniGard).
Green peach aphid (*Myzus persicae*)	Examine crops on a weekly basis. Found on a wide range of bedding and hanging basket plants.	Range in color from light green, light yellow, green, gray-green, pink to reddish. Have a pronounced indentation between the base of antennae in front of head.	See aphid section above.

(continued)

	Best method to monitor	Identification features	Potential biological control
Chrysanthemum aphid (*Macrosiphoniella sanborni*)	Found only on chrysanthemum species.	Insects are shiny, and color varies from reddish-brown to blackish-brown. Cornicles are short, stout, and black.	See aphid section above.
Melon aphid (*Aphis gossypii*)	Found on a wide range of bedding and hanging basket plants.	Color varies from light yellow to dark green. No indent between antennae such as green peach aphid has. Distinct cornicles that are always dark colored for the entire length.	See aphid section above.
Caterpillar (general)	Caterpillars of several species feed on greenhouse crops. Most adult butterflies and moths overwinter outdoors and migrate into a greenhouse. Regularly monitoring adult flight activity alerts you to when to look for eggs laid on foliage or stems.	Many caterpillars have appendages called prolegs on their abdomen. All caterpillars have chewing mouthparts they use to chew foliage and stems or bore into stems.	Using microscreening over vents and greenhouse openings can exclude migrating adult moths and butterflies. Use *Bacillus thuringiensis* (Bt) for early caterpillar stages. Bt is sold under several brand names.
Western flower thrips (*Frankliniella occidentalis*), flower thrips (*Frankliniella tritici*), onion thrips (*Thrips tabaci*)	Adults congregate in open flowers and can be easily tapped out of flowers. Many species tend to be found in tight, hidden parts of plants (notably western flower thrips), while others such as the flower thrips feed on open leaf surfaces. Feeding thrips deposit black fecal matter in circular shapes on leaf surfaces. Yellow sticky cards will capture thrips, but blue sticky cards are particularly attractive to thrips.	Adults are small in size, generally .04 to .08 inches long. Adult bodies are tubular with narrow, pointed, fringed wings. Two larval stages feed on plant parts above ground. Prepupal and pupal stages are in soil.	Using microscreening over vents and greenhouse openings can exclude migrating adult thrips. Predacious mites *Neoseiulus cucumeris* and *N. degenerans,* for first instar thrips larvae; minute pirate bugs, *Orius* spp., for larvae and adult thrips; the entomopathogenic fungus *Beauveria bassiana* as a fine mist spray.
Cuban laurel thrips (*Gynaikothrips ficorum*)	All stages—eggs, larvae, prepupae, pupae, and adults—found in curled leaves of tip growth of plants. Found attacking ficus, Indian laurel, India rubber plants, and several herbs.	Adults are black and tube-shaped with fringed wings.	Prune infested tip growth to reduce populations. Infested tip growth will often have sunken, reddish spots forming along the midrib vein. Microscreening can exclude migrating adults.

(continued)

	Best method to monitor	Identification features	Potential biological control
Silverleaf whitefly (*Bemisia argentifolia*), greenhouse whitefly (*Trialeurodes vaporariorum*)	Use sticky cards to monitor for adults. Place sticky cards near intake vents and doors to detect inward-migrating adults. Inspect plants to detect immature stages on under surfaces of foliage.	Adult whiteflies are small (.04 to .08 in.) in length and are white, flylike insects. Eggs are tiny, spindle shaped, and laid on under-sides of leaf surfaces. Silverleaf whitefly eggs start white but turn amber-brown. Greenhouse whitefly eggs start white and turn to gray with time. Crawlers and other nymphal stages are oval, flattened, and translucent.	For greenhouse whitefly, suppress using releases of *Encarsia formosa* early in a crop cycle before population buildup. For silverleaf whitefly, suppress using *Eretmocerus californicus* or the entomopathogenic fungus *B. bassiana* for nymph stages of silverleaf whitefly and greenhouse whitefly.
Fungus gnats (*Bradysia* spp.)	Sticky cards will capture adult fungus gnats. Lay potato slices (1 in. by 1 in.) on soil surfaces. Larvae will migrate to potato disk on the surface facing the soil media.	Adults are .16 to .20 inches long with dark brown to black bodies. Antennae are long and beaded. Wings have a distinct Y pattern. Larvae are long and slender with a black head capsule.	Entomopathogenic nematodes including *Steinernema feltiae* and *S. carpocapsae* for larval stages, *Bacillus thuringiensis* var. *israelensis* for larvae.
Mites			
Spider mite (*Tetranychus urticae*)	Mites tend to build up in hot conditions. Examine plants in the warmest section of the greenhouse for early infesta-tions. Look for stippling of foliage. In heavy infestations look for webbing on stems and flowers.	Larvae are pale green with six legs. Protonymphs and deutonymphs have eight legs and are pale green to brown. Adults have two large, black spots on each side. Adults have eight legs.	Predacious mites are *Phytoseiulus persimilis* and *Neoseiulus californicus*.
Diseases			
Pythium	Monitor plants that may have been stressed by high salt levels, excessive soil wetting, or high levels of fungus gnat larvae. Use Alert serological test kits.	Examine roots for cortex that sloughs off, leaving central core of root.	SoilGard (*Gliocladium virens*), G1-21, and Root Shield (*Trichoderma harzianum*).
Rhizoctonia	Examine bases of plants, especially for susceptible species such as begonias, impatiens, petunias, dahlias, and cyclamen. Use Alert serological test kits.	Examine for small, water-soaked spots on stems or leaves.	SoilGard (*Gliocladium virens*) G1-21. Root Shield (*Trichoderma harzianum*)

(continued)

	Best method to monitor	Identification features	Potential biological control
Botrytis blight **(Botrytis cinerea)**	Examine tender terminal growth, flowers, or lower leaves on crowded plants. Monitor closely during periods of cloud cover, cool temperatures, and when plants are crowded.	Look for tan to brown dead areas and gray fungal growth.	Improve air circulation around plants. Use horizontal air flow (HAF) fans to keep air circulating in greenhouses. Space plants to avoid overcrowding.
Powdery mildew **(Erysiphe spp.,** **Oidium spp.)**	Look for white or yellow spots on upper leaf surfaces. The disease is commonly found on poinsettias, gerberas, phlox, begonias, and chrysanthemums.	Monitor areas with poor air circulation and high humidity. Areas with large temperature fluctuations between day and night temperatures have higher incidences of powdery mildew.	Biorational materials such as horticultural oil (SunSpray Ultra-Fine Spray Oil and Stylet Oil).
Bacterial blight **(Xanthomonas** **pelargonii)**	Examine all geraniums closely during warm weather.	Look for isolated leaf wilting, wedge-shaped yellowing between veins, or eighth-inch round, brown spots. Also, look for vascular discoloration.	Ivy geraniums can be infected and be symptomless. Don't place ivy geranium baskets above geraniums. Remove and destroy all infected plants.

March 1997.

Sanitation: Key to IPM

Jim Willmott

Over the past growing season, you did everything possible to provide favorable temperatures, moisture, and nutrients for your plants. Unfortunately, this also favored an abundance of insects, mites, disease-causing pathogens, and weeds. Pests leftover from spring, or those present now in poinsettias and other holiday crops, are risks to your future profitability.

While there's been no shortage of integrated pest management (IPM) research, sermons, and literature, many growers still violate a basic principle of IPM: sanitation. While some growers make excuses that there's not enough time or that labor's too short, there's no legitimate excuse for dirty, pest-infested greenhouses. Remember, even though pests aren't evident today, they'll arise tomorrow as a more difficult and expensive problem.

Where Pests Hide

Pests are most obvious when attacking crops. However, this is not the only indicator of their presence. They survive and flourish on flower petals, foliage, and media that fall on benches, walkways, and floors. Foliage and flower debris harbors numerous key pathogens including *Alternaria, Botrytis, Xanthomonas,* and numerous leaf-spotting fungi. Debris may also be home to mites, thrips, aphids, whiteflies, and other insects. Crop media, roots, and excess irrigation liquid often contain pathogens such as *Pythium, Rhizoctonia,* and *Thielaviopsis.*

Once on the floor, conditions favor pest survival. Pathogens, for example, flourish in moist soil—especially when loaded with crop residue. Also, pathogens, insects, and mites survive and prosper on weeds. Breaking the link between accumulating crop debris and favorable conditions on greenhouse benches and floors is a critical IPM practice.

First, limit crop debris. Teach crop handlers the importance of grooming dying foliage and flowers. On registered crops, Florel can be used to prevent early flowering, a major source of debris. Make a habit of removing fallen debris from benches, walks, and floors. Sweeping floors is best done between crop cycles as it stirs pests, especially pathogens, into the air and into crops. Paved floors are easiest to clean by sweeping, using blowers, or washing with a hose. Gravel floors are more difficult to keep clean—consider paving or covering with ground cloth.

Next, clean up weeds. First, identify and understand their life cycles. Different types require different management. For example, in northern regions of the U.S., tender annuals can be controlled by allowing greenhouse temperatures to drop below freezing (be sure to follow proper precautions to prevent damage to irrigation and other production systems). Freezing, however, will not kill perennials or winter annuals.

Hairy bittercress and common chickweed are common winter annuals. Herbicides are effective, but only four are registered for use in greenhouses. All should be applied directly to weed foliage, and none have soil activity or prevent weed seed germination. Roundup is the only translocated herbicide. It's effective for perennial weeds, but it shouldn't be applied when crops are present. Finale, a recently labeled product, is effective on many weeds, including perennials, and can be applied during crop production. Two contact herbicides, Reward and Sharpshooter, are effective in killing annuals but not perennials. When applying Finale, Reward, or Sharpshooter, avoid crop contact. When crops are present, use low-pressure sprays with large droplet size and turn off air circulation fans. Always read and follow label directions when applying herbicides, and do not apply unregistered herbicides in greenhouses! Some, such as Casoron and Goal, are volatile and persistent and can result in significant crop loss.

The most effective, but costly, technique for minimizing weeds is hand removal.

Finally, use a disinfecting solution on greenhouse benches, equipment, and floors to kill infectious pathogens and algae. The most common products are sodium hypochlorite and quaternary ammonium salts. These have no residual contact, so treated areas must soak in solutions. Neither is effective on permeable surfaces such as gravel walkways or floors. Remove crop and media residues before soaking.

Sodium hypochlorite solutions are made by mixing one part of a bleach product such as Clorox (5.25% sodium hypochlorite) with nine parts water. It's important to use solutions promptly after mixing. Tools, equipment, and surfaces should soak for at least thirty minutes. Use caution because solutions are corrosive and can bleach clothing and irritate eyes and skin. Additionally, Cornell research reported injury to poinsettias and mums that were placed on recently treated capillary mats. They recommend rinsing and ventilation before setting crops on mats. Others have related injury to plants grown in treated plastic pots.

Quaternary ammonium salt products—including Greenshield, Triathlon, Physan 20, and others—are economical, effective, and have many advantages. They're more stable in solution than sodium hypochlorite and are safer for plants, equipment, and people. Apply to benches, walks, pots, flats, and other equipment. Generally, a ten-minute soak is effective, but longer periods are needed for algae on walkways. Routine applications during production are recommended to reduce the risk of crop loss.

Pest Control, October 1998.

Genetic Engineering: New Tools for IPM

Kevin M. Heinz and Jianbo Li

Genetic engineering has revolutionized the way we make medicines and grow field crops. It will also change floriculture production: Flower varieties will be genetically engineered to have enhanced postharvest longevity, novel shapes and colors, and resistance to pests and diseases, all of which could lead to significant profit increases for growers.

What Is Genetic Engineering?

Genetic engineering is a modern technique capable of altering the genetic code of a target animal or plant, resulting in what's known as a transgenic species. Transgenic crops can be produced by the transfer of genes from one species to another, whether or not they're related—a goal unachievable in conventional breeding programs. With recent advances in molecular biology, genetic engineering has routinely been done on crop plants and other organisms, such as fruit flies.

Many agricultural crops have been genetically engineered to provide resistance to pests, fungal or viral diseases, and other agronomic characteristics such as longer storage life of fruits and vegetables. For example, cotton varieties have been modified to produce their own pesticide against bollworms, potato and squash transformed to resist viral diseases, and tomato modified to have longer shelf life. More than twenty genetically modified field crops and vegetables have been approved since 1992 for large-scale field testing or commercial use.

The significant yield improvement of transgenic crops can have profound impact on agriculture. The use of transgenic crops that produce their own pesticides or fungicides will result in reduced use of chemical pesticides and lower residues in the environment. Techniques used to genetically modify those plants can also be applied to floral crops. Most flower crops—such as roses, carnations, tulips, and chrysanthemums—can be routinely transformed today.

Transgenic Floral Crops and Pest Management

Safe and economic alternatives to traditional pesticides have been sought and are being developed. Genetic engineering can be used on two pest control fronts: to produce transgenic crops that make their own pesticides and to transform microbial pesticides and natural enemies to make them more efficient.

Finding single genes that encode resistance to target pests has been the major challenge of genetic engineering in agriculture and floriculture today. Most of the agricultural crops that have been genetically engineered to make their own pesticides were produced by transferring a gene from the bacterium *Bacillus thuringiensis* (Bt), which produces insect-specific toxins that kill various moth larvae. Bt has been used as a microbial insecticide for about a quarter of century without any known adverse effect on human or other vertebrates. Genetic engineering has made Bt an even better pest control agent. Few other genes have been used because they are difficult to find and characterize.

Floral crops are currently being genetically modified to produce attractive shapes, colors, and fragrances; to increase postharvest longevity; and to reduce irrigation requirements necessary to maintain plant quality. For example, a gene from maize was introduced into petunia lines, and a bright-orange petunia variety was produced by conventional breeding and selection of these genetically transformed petunias. Carnation cultivars have been modified to produce low levels of ethylene, resulting in longer postharvest life. Floral crops are also being genetically modified to be resistant to pests. Commercial breeders have produced several lines of transgenic petunias and poinsettias with genes that may produce resistance to whiteflies. The effectiveness of these transgenic crops to manage pest problems is currently being evaluated. Genes encoded to kill pests (such as the Bt toxin gene), retard pest development, or repel pests from plants are all good candidates for gene transfer. Floral crops modified with these genes will provide us with a wide range of control measures for pest management.

Bioinsecticides and Natural Enemies

Genetic engineering has also been applied to biopesticides and natural enemies. Several major companies may soon market a genetically modified virus, the *Autographa californica* nuclear polyhedrosis virus (AcNPV). AcNPV is a naturally occurring virus whose major hosts are moth larvae such as loopers, armyworms, and bollworms. Virus formulation may be sprayed onto plants using conventional insecticide application technology. Pest larvae feeding on plant tissues ingest viral particles, and the virus kills its hosts as it reproduces within them. Although AcNPV occurs in nature, death to larvae feeding on this virus usually occurs after the insect has already done extensive crop damage. To have a fast-kill virus, a gene coding for an insect-specific toxin in scorpions has been introduced into AcNPV. The result is a genetically modified baculovirus that kills its insect hosts rapidly and reduces damage to foliage from feeding caterpillars by 50%. This genetically modified biopesticide is currently being tested in small-scale field trials.

Genetic control of aphids provides another interesting scenario. Because aphids feed by sucking fluids from their host plants, biopesticides such as AcNPV won't be effective since they require the pests to consume the virus lying on the surface of the foliage. However, insect-specific fungi readily attack aphids. These fungi could be genetically modified to increase their ability to infest aphids. Furthermore, transforming aphid-vectored plant diseases into aphid-killing and plant-benign diseases may also be developed in the future.

Researchers have been trying to genetically modify arthropod natural enemies for pesticide resistance and enhanced control efficacy. The progress in developing transgenic natural enemies, however, has been slowed by the difficulty in identifying and isolating appropriate genes. As more genes are identified, transgenic natural enemies could become available in the near future. We could modify parasitoid wasps to become better control agents of greenhouse whiteflies using the same genetic engineering techniques.

What Are the Risks?

Transgenic plants and animals may present risks in two ways. First, genetic modification may produce undesirable characteristics. Second, under rare circumstances the new gene, just like any other gene in a plant or animal, may be transmitted to other species. In nature, the genetic code of a species is transmitted from one generation to the next through normal reproduction, and the species identity is maintained. A gene may move from one species to another in one of the following two ways. Closely related species may cross, and some of them could produce viable offspring, resulting in the transfer of genes from one species to another. This kind of gene transfer could be completely avoided in greenhouse situations because close weedy relatives of floral crops could be excluded. In the other situation, genes may theoretically move between two closely associated species, such as a parasite and its host. Some examples of DNA pieces, not functional genes, are known to have transferred this way, possibly millions of years ago. In reality, such gene transfer is essentially irrelevant to transgenic floral crops or greenhouse pest control agents.

All transgenic crops and animals have been and will be evaluated under laboratory and controlled field conditions for undesirable characteristics before commercial use. Modification of one gene usually does not produce unintended characteristics. For example, no host range shift or change in egg production was observed in a genetically modified predatory mite. Even for those field crops in which hybridization with related species is possible, preventative measures can be found after careful evaluation. For example, a sixty-five-foot buffer zone of a non-transgenic cotton variety captured all dispersing pollen of a transgenic cotton variety. The risk of development of pest resistance to genetically engineered crops or other control agents is not unique to transgenic organisms, and it can be avoided or delayed with well-thought management strategies treating transgenic organisms in the same manner as conventional pesticides.

Genetic Engineering as Part of IPM

Genetically engineered floral crops and pest control agents have the potential to increase the productivity and quality of floral crops and significantly reduce the economic losses caused by pest damage. However, this technology alone will not be sufficient to solve greenhouse pest problems for these reasons: Most pests cannot be completely eradicated from the earth, and the continued presence of genetically engineered crops containing microbial insecticides or the preventative use of genetically modified biopesticides may lead to pest resistance, just as has occurred with the development of pest resistance to chemical pesticides. Therefore, the basic principles of IPM must still be followed when using genetically engineered crops and pest control agents. Transgenic crops and genetically modified pest control agents should be treated as individual IPM components within a floricultural production system, and they should be integrated with other currently available control measures to reduce pest problems below economic injury level. With less chemical residues put into the environment and more control measures available, we may realize a new level of pest control in greenhouse crops and enjoy more productivity and profitability.

Editor's note: The authors thank the American Floral Endowment for their funding of this project.

August 1997.

Chapter 2
Biological Controls

Biocontrol FAQs

Mike Cherim

Using biological control agents to control unwanted pests remains one of the more mysterious methods of pest control. We asked Mike Cherim, director of The Green Spot, a supplier of a wide range of biocontrol agents and IPM supplies, to share with us the most common questions he gets from growers who are curious about biocontrol.

Q: How difficult, on average, is biocontrol?

A: Not too bad. Perhaps it's on par with learning to work with some of our newest computing tools—though probably more fun and more reward-ing. Expect, like anything, a fairly steep learning curve at first, with a gradual leveling out. Do expect setbacks in the first few years. Your scout might have missed an area during scouting, and a problem now exists there. This isn't a reason to give up or get disgusted. After all, if folks gave up when chemicals failed them the first time, no one would be doing anything about pests except squishing them by hand today. In other words, biocontrol is no more difficult to grasp than controlling pests chemically, developing a marketing strategy, learning how to keep your books or even growing plants—desire is paramount.

Q: What's the most serious drawback to biocontrol?

A: The cost of product and shipping can be significant in many cases, but if labor is taken into account, the overall costs should be comparable to chemicals only.

Other drawbacks may also include the "hassle factor" of having to learn new tricks. This is especially bothersome to older dogs. And to compound their discomfort, implementing biocontrol can complicate their issues by reducing what is safely available for them to fall back on if they decide to spot-treat problem areas with some of their stand-by weapons, as many chemicals are not compatible with biologicals.

Q: How do I get started?

A: Find a good supplier to help with the transition and to help you make sure all your ducks are in a row. First things first: The sprays must stop. Not all sprays, though; just the incompatible ones with long residual periods. A good supplier can help you identify which sprays are OK for transition. Secondly, your crops should be relatively clean; when nothing else seems to be working is not the time to start a biocontrol program. Lastly, after you know, as specifically as possible, which pests you're dealing with or are expecting to, the groundwork for a good program can be laid.

Q: What should I look for in a biocontrol supplier?

A: The best bugs to use are those that are fresh, healthy, and ready to do their job. Not every supplier, though, offers the freshest possible product. And you may not be able to tell just by looking. Other criteria will make it easier for you to have faith in your supplier. For example: Expect a delay of about one week from the time you place your order. Since biocontrols are living things, they can't normally be placed on a shelf and shipped immediately at your request. The good bugs must first be collected, sorted, etc. This takes time.

Another criterion is the shipping method. For example: If a supplier is willing to ship by a means other than one that will result in delivery within 24 hours, he's not really doing you a service. Your biocontrol supplier should not only be able to give you the most up-to-date release strategies, chemical compatibility information, and other technical support, they should be poised to work in your best interests, no matter how hard you kick or how loudly you scream.

Q: In scouting for pests in a chemically treated facility, the scout looks for dead bugs to see if a chemical application has been effective. How does this differ in a biocontrol program?

A: Pests affected by natural enemies in a biocontrol program may not be dead by conventional standards. In many cases, they will be as good as dead but not actually so. In such cases, the pests may be parasitized, with guaranteed doom impending. Confirmation is usually visually detectable. In many cases, the only really reliable method of determining whether or not your pests are being controlled is to scout the crop's new growth to see if it has emerged pest free. Your biocontrol supplier should be able to give more specifics as to what to look for, where, how, and when.

Q: What should I do if I have a lot of time and money invested in biocontrol, yet things are beginning to go awry?

A: The key word is beginning. Don't wait until things are in serious disorder. Based on information gleaned from your thorough scouting, trends should be detectable and correctable early on in your program's decline.

Now, to turn things around without resorting to the "big guns" (which, incidentally, are reported to work better than ever after using biocontrols for a period of time), you should consider using some of the softer, more compatible biorational substances to treat flare-ups and hot spots. Again, this is another case in which your biocontrol supplier should be able to help.

Q: Biorational substances: What are they? What can I spot-treat with?

A: Certain chemicals are taboo because they will kill your biocontrol agents outright (your supplier can give you the specific names of these chemicals). Others are OK to use just before you start, as they aren't compatible but have a very short residual period. Yet others are OK to use while the good bugs are working. Spot treatments should be made with the latter two categories only. And special emphasis should be given to the use of only compatible products where possible.

Q: Will biocontrols work in my situation?

A: Probably yes. Long-term crops, short-term crops, indoors in greenhouses, outside in fields and orchards, interiorscapes, research facilities—the list goes on and on. Biocontrols won't work in every conceivable situation, but they've been successfully used in most situations. Again, this is something best discussed with your chosen supplier.

Q: What can I expect of my plant sales if I allow a few pests to survive?

A: Almost all consumers have heard how pesticides are supposed to be less than perfect for our world. So you can capitalize on this by marketing your wares as having been treated with fewer chemicals (if your plants are healthy). Some growers have made names for themselves by "going public" with their new pest control techniques. They've hosted tours, given interviews, and have been, basically, at the center of their local limelight. This is good for business. Of course, you must initiate these events at first, but once word gets out . . . well, we know of one nurseryman who is now considered somewhat of a celebrity. For him, all avenues of biocontrol work exceedingly well.

April 2000.

What'd He Say?

As with every technology, biological control has its own specific terminology. Here are a few key terms and their definitions from the article "Biocontrol FAQs" so that you and your staff will speak the same lingo as your biocontrol supplier.

Biocontrol—A new compound word for biological pest control. To practice biocontrol is to use living, multicellular organisms (biocontrol agents) to hinder or eliminate plant pests.

Biorational—An item or substance thought to be generally compatible* with biocontrol agents and thus useful as part of an IPM program. This usually pertains to certain pesticides (i.e., insecticides, acaricides, fungicides, etc.). Also called a "soft chemical."

Parasitoid—An insect, usually a wasp, that lays eggs inside a host organism (a plant pest in this case). The larva develops inside the host and eventually kills it. This is different from a parasite, which will live on or with a host organism but doesn't kill it.

Predator—An organism that preys on other living organisms (plant pests in this case).

Compatible means that the substance causes less than 25% mortality to biocontrol agents within twenty-four hours after direct contact.

Culture Notes, April 2000.

Adopting Biocontrol into Pest Management

Sherri Bruhn

The organic gardener's mantra is that the fewer chemicals you use, the fewer long-term environmental side effects will occur. Unfortunately, it isn't that simple. While all is well and good with small-scale organic growing, those who make plants their livelihood know it isn't necessarily a viable answer. As a compromise, some growers have incorporated biological controls alongside their chemical regimens. They've seen successes and failures along the way, but within their experiences lie valuable lessons about biocontrol agents.

Defining Biocontrol

According to Dr. Kevin Heinz, associate professor of entomology at Texas A&M University, College Station, Texas, biological control is essentially using living natural enemies to suppress pest populations. This can occur in a number of ways: Parasitic enemies such as wasps can use pests as hosts for their offspring, predators can consume pests as prey, or pathogens can be introduced to cause disease within the population.

The trick is that for every pest there can be a different predator, and sometimes even the plant cultivar can help or hinder the entire process. This leads to an infinite number of relationships and possibilities. No two greenhouses are alike in their crops or in the pests that plague them. For this reason, Kevin says the end goal isn't to eliminate chemical spraying, which is unrealistic, but to decrease it and use chemicals as a part of an overall IPM strategy.

The Softer Bounce

Barker Ranch, Atascosa, Texas, is part of the Color Spot Nurseries network, and Chris Fox, technical services leader, will tell you that he's always looking for biological alternatives to stop the pests that infest Barker's purple sage, hibiscus, *Sophora,* and other crops.

"The whitefly took me to school and taught me that the harder you hit them, the harder they hit you back," he says.

In order to minimize pest resistance, Barker Ranch's strategy is the "soft bounce." According to Chris, insects don't become resistant to biologicals, as they do to neurotoxins. Previously, he used neurotoxins to control whiteflies and realized that he had inherited a very resistant strain. After trying many neurotoxic products, he didn't see the results he was looking for. Chris reports he's had good experiences for eight years with BotaniGard and other beneficial fungal products on the market. Once something matches up well against a certain pest with a particular crop, he says the results are well worth the effort.

Other biologicals he recommends are Dipel and Conserve for worms, and bacteria-based Mosquito Dunks minimize mosquito traffic from standing water sources. The donut-shaped dunks control larvae for three to four weeks and are useful in the warm parts of the Texan year.

Barker Ranch's team integrates biologicals with insect growth regulators, neurotoxins, herbicides, fungicides, repellents, and fertilizers. They base application decisions on scouting and significant populations, primarily for economic reasons. When Chris does apply biologicals, he uses sprays or drenches according to the labels' application rates, and he says the payoff is long term.

"When dealing with biological systems," Chris explains, "there aren't a lot of hard and fast answers. The results are qualified. We do a lot of trials to find out what [each product] works well on."

Planning a Schedule

Whether you're using biologicals regularly or using them on an as-needed basis, long-term economics point to spiraling decreases in pest management costs.

At Zylstra Greenhouses, a 200,000-square-foot bedding plant facility in Kalamazoo, Michigan, the total cost of spreading *Aphidius colemani* (parasitic wasps) is less than spraying a couple of acres for aphids just once when factoring in labor and chemical costs. Owner Steve Zylstra says, "Now, we just monitor and spot spray."

Steve's routine for combating aphids is to spread *Aphidius* around at a rate of approximately five hundred larvae per 20,000 square feet over plugs and cuttings. The larvae come in a shaker bottle, and Steve distributes them into empty hanging baskets every two weeks from January to March. He makes sure they stay dry during the week or so it takes them to hatch. Then they fly out and parasitize aphids with their eggs. Even though initial trials were in two hoop houses, Steve says *Aphidius* were all over the farm after they'd hatched. He's been using beneficials for the past five years and hasn't seen a problematic aphid infestation since.

That doesn't make beneficials foolproof. It's important to release *Aphidius* before aphid populations are too heavy. If the populations are too high, the wasps can't catch up. It's also necessary to keep weeds down inside the greenhouse. And when you apply sprays, always keep your good bugs in mind.

"You have to understand the nature and life cycle of the beneficial as well as the target insect. We don't want to eliminate our beneficials," Steve says. "For some things, you have to be careful not to use a spray with residuals. We understand what sprays will work. We use Azatin halfway [one week] in between the release of Thripex, not one or two days after the release."

Being Culturally Sensitive

Jim Atchison, owner of Atchison Exotics Inc., Delray Beach, Florida, hopes to become certified as an organic grower. He applies herbicides at half rates and uses Dursban against fire ants when he ships to quarantined areas.

Trials and Tribulations: A Biocontrol Calendar

Steve Zylstra trials many pest controls to see what works and what doesn't. By keeping detailed records of what he uses and when he uses it, he can accurately evaluate the success or failure of a particular program. Here's an example of his 1998 calendar:

January 14—Applied Thripex to control thrips on young plugs and cuttings. Used bottle of 25,000 mites per 4,000-sq. ft. bay, four bays on a weekly basis. Cost: $20 per bottle.

Follow-up: Weekly scouting with blue sticky cards. Azatin sprayed two days before next release if two to three thrips were found on the cards.

February 26—Began releasing *Aphidius colemani* over plugs, cuttings and transplanting area. Released 500 per 20,000 sq. ft. on a weekly basis. Cost: $20 per release.

Follow-up: Weekly scouting. Spot sprays for aphids.

March 1—Applied Thripex to control thrips on potted New Guinea impatiens, gerbera daisy, etc. Used 25,000 mites per 4,000 sq. ft., within a 15,000-sq. ft. area.

April 20—Developed a whitefly problem. Aggressively sprayed to control it and stopped releasing Thripex.

June—*Aphidius* activity still evident in crops.

Evaluation: Control with Thripex justified the cost in a high-dollar crop, but as Zylstra Greenhouses began brokering plant material, Steve didn't want to risk a surprise infestation. He'll review the program again next season.

"I don't blanket spray anything," Jim says. "[Blanket spraying] is just the mentality: 'If it's moving or crawling, it's bad.'"

Jim estimates that he spent $4,400 on predators for 1998. His main controls are the native predators already present at his facilities. He uses three main predators he buys as supplements, and uses them in his shade houses on schefflera and *Ficus lyrata*. *Phytoseiulus persimilis* mites are broadcast to keep down two-spotted spider mites, while green lacewings and pink spotted beetles keep down aphid populations and control mealybugs somewhat.

Release is a critical part of the planning. Some predatory insects will eat one another if given the chance, so Jim separates the cannibalistic lacewings as soon as possible, before the larvae mature. He uses rice hulls to apply them to schefflera leaves. Jim releases five thousand lacewings with five hundred pink spotted beetles, one per plant, every two weeks to keep the upper hand.

What does he do for thrips control? He takes a cultural approach and only grows plants that can be grown as organically as possible.

"I avoid growing varieties that are problematic, things that I can't fit into the program," says Jim. "I grow plants that are culturally correct, and I keep the weeds down." In order to keep the plants looking clean, Jim takes off leaves with heavy populations before plants are shipped.

For growers who want to start a biological program, Jim recommends starting with a clean nursery and working with an entomologist to find out

Green lacewing larvae are distributed individually with rice hulls on the upper leaves of Atchison Exotics' schefflera.
Photo by Sherri Bruhn.

how chemicals and beneficials work together. He also stresses the importance of sticking with the program because it can take a long time to balance out.

Patience Is a Virtue

Danny Giesbrecht uses *Orius* (minute pirate bug) to control thrips at Lake Shore Flowers, Leamington, Ontario, which has 130,680 square feet of alstroemeria. He says that out of the many biological controls he's tried, the *Orius* predator is one of the only things that's worked to keep thrips counts down. He releases *Orius* every week, beginning in the last week of March and continuing for eight to ten weeks afterward. He estimates that he releases an average of 60,000 *Orius* per year, at a total cost of $3,168 (about a nickel per bug). Thrips have been absent on sticky cards since early November 1998. Danny also credits naturally occurring pests for the decrease in other pests' numbers. And, the less he sprays, the fewer beneficials he kills.

"We haven't sprayed in three years, with the exception of spot sprays for aphids," he explains. "Biologicals are now coming in naturally. The further we get away from [relying on sprays], the more we're finding that our thrips numbers are going down."

It's essential to be aware of the relationships between biologicals and chemicals when using them together. One of the biggest mistakes growers make when using beneficial pests is using chemicals in nearby areas. While pesticides will knock out problematic pests or diseases in one crop, they may also be killing both naturally occurring and store-bought predators nearby. Not only does this waste money, it also makes it impossible to determine if the beneficials are doing their jobs.

"The first three years were hair-pulling because we played around with different things," Danny says. "It's difficult to get started, but once you get success in a crop, there's no better way. It takes lots of patience, but the more people you talk to, the better off you are."

Forecasting Things to Come

Graeme Murphy, pest management specialist for greenhouse flowers with the Ontario Ministry of Agriculture and Food, Vineland Station, believes there's a future for biological control. "I think the interest level in biological control is only going to increase," Graeme says, "but there will always be a need for pesticides. In fact, many new pesticides on the market are being developed with a high compatibility with biological control programs."

Graeme asserts this trend will support biological controls and that it will work to prolong the life of pesticides. Resistance management will be an additional benefit as growers reduce the volume and rate of chemical applications.

Author's note: For a comprehensive guide for identifying pests and their natural enemies, try the Ball Identification Guide to Greenhouse Pests and Beneficials *by Stanton Gill and John Sanderson (Ball Publishing, 1998).*

April 1999.

Biological Control Experiences—
An Ontario Perspective

Graeme Murphy

Biological control in the Ontario greenhouse industry is gaining in popularity, although it's very difficult to put a figure on the extent of its use.

A survey of the Ontario greenhouse industry in 1994 showed that 39% of growers had used biological control at some time, and there is no doubt that this number has increased substantially since then. As the use of biocontrol increases, we're starting to better appreciate the situations where it will be most effective.

Aphids

I often comment that if aphids were the only pests growers had to deal with, everyone would use biocontrol. That may be a slight exaggeration, but there is no question that *Aphidius colemani* (the most commonly used natural enemy against aphids) is very effective.

This wasp controls both the green peach aphid and melon aphid, which are the most common aphid pests in Ontario. It also survives a moderate frequency of pesticide treatments because it's protected in the parasitized aphid. It probably isn't a good idea to rely on this ability to survive chemical warfare, but I've seen *Aphidius* remain effective in greenhouses for several years where pesticides are being used. It's very mobile and has a wonderful ability to spread from a small release area to the rest of the greenhouse. And compatible pesticides are available if things do get out of hand. On the downside, we've seen *Aphidius* succumb to hyperparasitism (parasites of parasites), which can reduce its effectiveness.

Spider Mites

In Ontario, spider mites are mainly sporadic, with exceptions in crops such as roses, ivy and hibiscus, where chronic, long-term infestations are often the rule. Little biocontrol has been used in roses, but in other crops, the success rate has been very good.

The most common predator is the mite *Phytoseiulus persimilis*, whose use pattern is best described as a biological pesticide. It's applied when mites are found and usually controls the pest so well that the predator eventually starves. When new infestations are seen, you introduce again. This works very well, but monitoring is obviously a key component in detecting infestations before they cause too much damage.

Thrips

Western flower thrips is the major stumbling block for biocontrol in Ontario (as it is elsewhere). However, growers have reported successes, many of which have a number of things in common: The most successful control begins when pesticides fail, as growers realize that pesticides aren't an option. Successful growers always have a strong commitment to making the program work. And, as time progresses, the grower's attitude and philosophy toward pest management changes. (With pests such as aphids and mites, results are quicker and more gratifying for the grower. With thrips control, you're in it for the long haul.)

The predatory mite *Amblyseius* (=*Neoseiulus*) *cucumeris* has been used successfully when introduced regularly before thrips numbers increase. Humidity is important for this mite, and we've seen it perform more effectively in double-poly greenhouses than in glass greenhouses. *A. cucumeris* is very difficult to find on crops after release, and this can be frustrating.

The minute pirate bug *Orius* is also used but has shown mixed results. It needs high thrips numbers or another food supply such as pollen, but even then, establishment has been inconsistent. When it gets established, control has been effective.

Many growers here also use the predatory mite *Hypoaspis miles* to help control thrips pupae in the soil. It can provide supplementary control, but it's difficult to determine its impact compared to other natural enemies used.

Fungus Gnats

Fungus gnats are the most common pest target for biocontrol. Hypoaspis, if used consistently over a period of months, has been effective. For more immediate results, the nematode *Steinernema feltiae* provides faster control than *Hypoaspis*. It has become a common treatment for fungus gnats, even for growers who reject the use of other natural enemies. There are still some concerns about its consistency and how to monitor for the presence of the nematode.

Graeme Murphy is greenhouse crops IPM specialist, Ontario Ministry of Agriculture, Food and Rural Affairs, Vineland, Ontario.

Pest Control, August 1997.

Two Biological Products Gain Popularity

We don't have much experience with the product, but enough soil manufacturers are now offering to add Natural Industry's Actino-Iron to their soil mixes to make it worth looking at as a soil additive. Natural Industries says that Sun Gro Horticulture is the latest to offer their product as an optional amendment, joining Al-Par Peat, Berger Peat Moss, Fafard, Florida Potting Soils, Michigan Grower Products, Phillips Soil Products, Pro Gro, and West Creek Farms.

What is this Actino-Iron stuff? According to the manufacturer, it's a combination of iron fulvate and a beneficial soil microorganism called *Streptomyces*. The *Streptomyces* colonize on the roots and secrete protective enzymes that help protect the roots from diseases. The microorganisms also break down organic matter and help release the iron and other minerals so they're more readily available to the plant. The effects last for one growing season. It can be applied as a drench or mixed dry into the soil before potting. Best of all, it's safe to handle.

Testimonials from growers who've tried the product look impressive, with one Florida foliage grower raving that his nandina crop finished in four months instead of twelve (the manufacturer says that because it improves root growth, it can cut crop times by 20 to 30%). Another foliage grower cut liquid feeds from five times a week to two and cut disease losses from 5 to 7% to less than 1.5%. A poinsettia grower who tried it on 20,000 plants says he only lost three plants total and didn't use any fungicides.

Another relatively new product that's attracting grower attention is Rootshield from BioWorks Inc. It's a strain of the organism *Trichoderma harzianum* that's a naturally occurring biological fungicide that also protects against root diseases. A major Illinois grower says they've seen improved growth on pansies, Christmas cactus, dracaena spikes, and poinsettias. And a large, multi-location grower based in South Carolina says that after testing it, they now use it regularly at all their locations for disease prevention.

While we avoid manufacturers' claims and promises, we pay attention when growers are saying good things about products, and we figure it's worth passing their comments along so that you can try the products for yourselves. Contact Natural Industries, Houston, Texas, Tel: (281) 580-1643, and BioWorks, Geneva, New York, Tel: (800) 877-9443.

Culture Notes, February 2000.

Testing Compatibility of Biologicals and Chemicals

*Michael P. Parrella, Dave vonDamm-Kattari,
Gina K. vonDamm-Kattari, and Pablo Bielza Lino*

While biological control for pest control in greenhouses is gaining momentum, it's still a painfully slow process for those of us who are committed to its principles and concepts and believe it will one day play an important role in greenhouse pest control.

One major hurdle is that insecticides still dominate the greenhouse. Certainly, biological control by itself is unlikely to ever provide the level of control that growers and consumers of ornamental plants demand. While there are examples where biological control is the dominant (and in some cases the only) pest control strategy employed, these cases are the rare exception rather than the rule. For biological control to continue to make inroads into floriculture, it must be integrated into an overall IPM program that includes the use of pesticides.

Pesticides and natural enemies are rarely compatible—the use of one usually precludes the use of the other. This has certainly been true historically, given the broad-spectrum biocides that growers commonly used. With few exceptions, the organophosphate, carbamate, and pyrethroid insecticides available over the past thirty years aren't compatible with natural enemies. But as these pesticides are replaced with reduced-risk materials, there's a good deal of excitement over the fact that they may be used in harmony with natural enemies.

Could the advent of reduced-risk pesticides allow the expansion of biological control into floriculture greenhouses? In some cases, chemical manufacturers have capitalized on this by touting that their materials are compatible with natural enemies or fit into an overall IPM program. However, details on which natural enemies are compatible and even what compatibility means are difficult to find. In addition, specifics of what kind of IPM program their material fits into are often vague and, from a practical sense, useless. Are there reduced-risk pesticides that are compatible with natural enemies? With funding from the American Floral Endowment, we've been asking this question over the past several years, testing several reduced-risk products and a variety of natural enemies. The following is what we've found.

What Does Compatibility Mean?

At first glance, the issue of compatibility may seem simple (i.e., does the chemical kill the natural enemy or not?), but in reality it's quite complex. There are at least five ways to approach the compatibility of natural enemies with pesticides:

Direct contact—If the natural enemy is directly contacted with the sprayed material, is it killed? For parasitoids that develop internally in their host (for example, aphid parasitoids), does the material kill these developing natural enemies?

Residue—For how many days after spraying is the residue of the material toxic to the natural enemy? This is very important in the greenhouse where repeated releases of natural enemies are often made.

Repellency—Perhaps the residue of the product doesn't kill the natural enemy but repels it. This could take the natural enemy out of the system and could be just as devastating to successful biological control as if the material had killed the natural enemy.

Sublethal effects—With some reduced-risk pesticides, such as insect growth regulators, there's the possibility that when applied to immature stages of a natural enemy (such as the aphid parasitoids mentioned above) they won't kill the natural enemy but could affect its ability to reproduce and lay eggs. Because of the short-term nature of most floriculture crops and the low tolerance for pests and pest damage, we don't usually expect the offspring of the natural enemies we've released to provide control. Still, the concern is that some of these pesticides, through direct contact or contact with the residue, may alter the behavior of the natural enemy such that it's not as effective at finding and killing hosts.

For example, suppose we have an insecticide that kills 50% of the natural enemies within forty-eight hours of contact. Is this material compatible because it only killed 50%? What about the other 50% that was contacted by the pesticide but wasn't killed? It may be unrealistic to expect these natural enemies to be unaffected by the spray. We expect them to move through the crop and find and kill pests in their natural way, but will they?

Another possibility is that natural enemies may not come into direct contact with the pesticides but may feed on pests that have come into contact with the material. This is quite common when soil-applied systemic materials are used. A foliar-dwelling natural enemy may not come into contact with a material applied to the soil, but it may pick it up as the pesticide travels throughout the plant and into aphids or whiteflies that are feeding on the treated plant.

Host elimination—Natural enemies must have prey available to survive in the greenhouse where they're released. If a pesticide reduces the pest population to such a low level that natural enemies can't find pests, they'll either die or leave the greenhouse. Such a pesticide isn't compatible with the natural enemy because it eliminates the host necessary for natural enemy survival. Granted, this is sort of an oxymoron because the overall goal of the control program is to drop pest numbers to near undetectable levels. The bottom line is that it would be a waste of time to use a natural enemy in conjunction with a pesticide that's so effective against the target pest.

Two points can be made here: First, many of the reduced-risk pesticides available don't approach 100% control after a single application. Second, the pesticide may target one pest and the natural enemy another.

To amplify the latter point, you may have a pesticide that works effectively on aphids, virtually wiping them out, but this may be in a greenhouse where you're using predatory mites to control spider mites. As long as the material isn't toxic to the predatory mites, the material may be compatible with biological control of mites—but it isn't compatible with biological control of aphids. A real-world example of this is the material Pirimor, which was used effectively as an aphidicide in greenhouses in Europe for many years (prior to resistance problems) and was compatible with biological control of mites and whiteflies.

Many other factors can affect the compatibility of pesticides and natural enemies, including rate applied, number of applications, degree of coverage, stage and age of the natural enemy present, timing of the pesticide application, environmental conditions in the greenhouse, and host plants treated.

It's easy to see how it can take years to fully evaluate the compatibility of a pesticide with a single natural enemy. The tedious nature of this work often precludes it from being done, and ultimately you want an evaluation of the material in the field (at the population level) to see if the product is compatible in a real-world scenario. We'll talk more about this later.

Where to Start?

In my laboratory, we try to keep things as simple as possible right from the start. We try to use the highest recommended label rate of the pesticide in question and evaluate direct contact and/or residual activity using adults of natural enemies we raise ourselves or purchase commercially. The age of the natural enemies are standardized to the best of our ability.

We use a laboratory spray tower to apply a known amount of material per unit area and either directly treat natural enemies on leaves or the leaves themselves, adding the natural enemies when the foliage dries. Chrysanthemum and rose leaves are typical test plant material.

Pesticides and Natural Enemies We Evaluated

Pesticide trade name	Chemical name	Rate[1]	Natural enemies evaluated
Avid	Abamectin	4 oz.	OI[2], PP[3], AC[4]
Alpha Hexyl	Alpha hexyl cinnamic	0.25%	CC[5]
BotaniGard	*Beauveria bassiana*	1 lb.	CC
Cinnamite	Cinnamic aldehyde	85-126 oz.	OI, PP, DI[6], AC, AA[7], CC
Conserve	Spinosyn	22 oz.	OI, PP, DI
Floramite	Bifenazate	4 oz.	OI, PP, CC, AC
Mesurol	Methiocarb	2 lb.	OI, PP
Orthene	Acephate	10.5 oz.	AC, AA, CC
Relay	Pyridine azomethine	5 oz.	OI, PP, AC, AA
Sucrose octanoate	Sucrose esters of fatty acids	0.15%	CC

[1]Formulated product per 100 gal. water   [4]*Aphidius colemani*   [7]*Aphidoletes aphidomyza*
[2]*Orius insidiosus*   [5]*Chrysoperla carnea*
[3]*Phytoseiulus persimilis*   [6]*Diglyphus isaea*

After treatment, leaves are placed in small, ventilated cardboard containers or petri dishes and held at room temperature. Mortality readings are usually taken after forty-eight hours and may extend for up to a week. Experiments are always replicated, and a water-only control is included as a treatment. To give an idea of numbers used, the trial involving *Orius* used twenty-five adults per container and was replicated five times for each pesticide evaluated. Pesticides we decided to evaluate were generally newer reduced-risk materials, but older, more established products were often included for comparative purposes (see above table). We chose commercially available natural enemies based on their potential to control major floriculture pests. We continue to collect data and have amassed a substantial amount of information. Here's just a portion:

Results

Floramite, Cinnamite, and Relay were relatively nontoxic to the generalist predator *Orius* in direct contact sprays when the mortality readings were taken forty-eight hours after application. There was no difference between the mortality caused by these materials and that caused by the water-only control. In contrast, Avid, Mesurol, and Conserve were toxic to *Orius* adults.

Tests with the leafminer parasitoid *Diglyphus* yielded strikingly different results when Cinnamite was compared to Conserve; Cinnamite was relatively nontoxic to adult parasitoids. Up to this point, it appears that Cinnamite and Relay exhibit good compatibility with natural enemies. However, it's dangerous to generalize, as natural enemies can vary greatly in how they respond to pesticides. For example, only Floramite proved to be compatible with the aphid parasitoid *Aphidius,* and no material evaluated was compatible with the aphid predator *Aphidoletes.* Against immature stages of the generalist predator *Chrysoperla,* only Floramite proved to be compatible. A final trial involved the compatibility of Floramite with the predatory mite *Phytoseiulus persimilis.* Floramite proved to be very consistent; its compatibility with the predatory mite was remarkable.

Lab versus Real World

As mentioned earlier, there are many factors that can affect the compatibility of pesticides with natural enemies. However, simple laboratory tests are a good place to start the evaluation process. The direct contact and residual bioassays put the natural enemy into a worst-case scenario where they're directly sprayed or forced to remain either on or in close proximity to freshly treated residue. In a field situation, it would be rare to directly contact all natural enemies present in the crop. Plus, spray coverage wouldn't be as thorough, especially on the undersides of leaves, which could allow natural enemies to move about without contacting fresh residue. In addition, if applications are made outdoors, ultraviolet light or other environmental factors could breakdown the residue, yielding greater levels of compatibility than were found in our laboratory. For a product such as BotaniGard, the insects were put into an environment highly favorable for survival and sporulation of the fungus—such conditions would rarely be met in a field or greenhouse situation.

The studies presented here help guide what the next level of experiments should be as we move toward making a more definitive statement on compatibility. With what we've seen so far, we can suggest which products look good, but we won't rule out materials that didn't perform as well in our tests. While Floramite looks like a very promising material, studies haven't been done on repellency, sublethal effects, or prey elimination. Products such as BotaniGard and Conserve need to undergo additional evaluations with real-world greenhouse conditions. Ultimately, the final evaluation of a potentially compatible material needs to be done in the field.

November 1999.

Safely Using Chemicals with Beneficials

David Cappaert

Faced with the inconvenience of reentry restrictions, growing costs, and an increase in local environmental regulations, you've made a commitment to biological control on your bedding plant crop. For ten weeks, you've used predatory mites to control fungus gnats and thrips. Parasitic wasps have kept aphid numbers low.

But three weeks before the shipping date, your pest scout reports aphid colonies on susceptible varieties and recommends spraying. Did your biological control program fail? Will you lose the control capacity of natural enemies in the crop if you have to spray?

The short answer to both questions is no, not necessarily. You can deal with the conflicting strategies of chemical and biological control if you have reasonable expectations and a knowledge-based strategy.

Chemical control has a direct and satisfying effect: the immediate destruction of damaging pests. Combined with regular scouting, a chemical control program allows a grower to spend money on pest control only as thresholds are exceeded. The disadvantages of chemical management are the indirect effects: phytotoxicity risks, elimination of natural enemies, unsightly and sometimes toxic residues, and the risk of resistance.

On the other hand, effective biological control begins before pests can reach high levels. It often depends on less evident indirect effects: preventative kill of early pest generations and a buildup of natural enemies in the crop. The disadvantage of biological control is its variability. Balancing the timing and numbers of introductions with environmental conditions can be difficult. Even a very good biological control program may require corrections with pesticides. The challenge in combining these two strategies is in preserving the built-in control of biologicals while minimizing the indirect effects of pesticides.

Reconsider Spray Thresholds

Growers often spray long before pests reach damaging levels. This is a prudent practice—if a grower knows from experience that ten thrips on a sticky card will grow to thirty in a week, he'll spray today. However, if he has a healthy population of *Amblyseius cucumeris* (a predatory mite), the thrips population will grow more slowly or may decline over time. Similarly, a whitefly population that seems to require treatment may be already under control if parasitism by *Encarsia* is 90%. Spraying in such circumstances may actually result in future loss of control by eliminating beneficials and their food source. Recalculating thresholds takes experience, good advice, and some courage—experiment at first in small areas.

Choose the Right Chemical, Application

A single application of many standard pesticides will end a biological control program for weeks or months. At the same time, there may be a chemical strategy with equal effectiveness that will allow beneficials to thrive. What's the difference? The chemical you choose, your application technique, and your timing are all important.

The most compatible pesticides are selective—much more toxic to pests than to beneficials. The clearest case is for the bacterial toxin *Bacillus thuringiensis* (Bt). There are formulations of Bt that are specific to Diptera (Gnatrol for fungus gnats) or Lepidoptera (Dipel for caterpillars). Bt has little or no direct impact on most beneficials.

Insect growth regulators (IGRs) are generally friendly to beneficials. They target immature stages, so they have little effect on adult beneficials. In many cases, they're also selective for pest species. Examples include azadirachtin (Azatin), derived from the neem tree; kinoprene (Enstar II), which is selective for Homoptera (aphids, whiteflies); and diflubenzuron (Adept), selective for Diptera (leafminers, fungus gnats, shore flies) and Lepidoptera (caterpillars).

Generally, chemicals that act as nerve toxins—organophosphates, carbamates, and pyrethroids—are toxic to natural enemies. Many have long-term residual activity, so that foliar treatments with malathion, bifenthrin, or chlorpyrifos will interfere with most beneficials for many weeks. This is especially important to understand if you buy in cuttings. Difficulty with biological control can often result from cuttings that have pesticide residue on the foliage.

Timing

Another strategy for minimizing impact on beneficials is to be selective about the timing and scope of pesticide applications. Nonselective pesticides with very short residuals can be used before introducing beneficials or in hot spots. Horticultural oils and insecticidal soaps are good choices for suppressing soft-bodied pests (aphids, whiteflies, and mites). They'll kill many beneficials on direct contact, but they won't harm beneficials after the spray dries. This allows

Strategies for Combining Chemicals and Biologicals

Strategy	Examples
Improve biological control	Add new species to control minor outbreaks—for example, ladybeetles to augment aphidius for aphid control
Use biological pesticides	*Bacillus thuringensis* for loopers; *Beauveria bassiana* (Botanigard) for whiteflies
Use insect growth regulators (IGRs)	Azadirachtin, Enstar II, Kinoprene
Spray when beneficials are less vulnerable	Nonpersistent pesticide (e.g., resmethrin) applied when whitefly parasites protected inside host
Focus on pest concentrations	Spray aphid hot spots with soap/azadirachtin; direct spray to whitefly colonies on new plant growth

the option of spraying individual plants that host aphid or whitefly colonies, while leaving beneficials on adjacent plants as sources of recolonization.

It's important to note here that not all beneficials are equally at risk. For instance, parasitic wasps such as *Encarsia formosa* (for whitefly control) are among the most sensitive. Predatory mites have a wide range of susceptibility—some strains have been selected for pesticide resistance.

In other cases, a biological is protected by its immediate environment. Parasites are much less susceptible when developing inside their hosts, and soil-dwelling predatory mites are quite resistant to foliar sprays. Nematodes are the most resistant beneficials, compatible with many broad-spectrum insecticides that would decimate most beneficial insects. How can you know what effect a pesticide will have on a particular biological control? Ask your biological control supplier—he should have practical experience and an interest in careful choices.

Biological control is rarely the only tool needed, especially in the floriculture industry, with its low damage tolerance. But with the new selective chemicals available, you can combine biologicals with pesticides and reduce sprays while gaining experience in the use of natural enemies.

Pest Control, June 1998.

Improve Your Bt-ing Average

Stanton Gill

Microbial insecticides, such as good old Bt—*Bacillus thuringiensis*—are often perceived as being less effective or slower acting than their chemical counterparts. In truth, these bacterial pathogens are safe to workers and deadly to greenhouse pests such as caterpillars, fungus gnat larvae, and leaf beetles, making them essential to every grower's IPM program.

Bacterial pathogens used for insect control are spore-forming, rod-shaped bacteria in the genus *Bacillus*. Most insecticidal strains of *Bacillus* have been isolated from soil. The most commonly used microbial insecticide since the early 1960s are formulations of Bt. The bacterium produce a spore and a crystalline protein toxin called an endotoxin. Most commercial Bt formulations contain the protein toxin and spores, but some are cultured in a manner that yield only the toxin component. The insecticidal activity of Bt derives from the protein crystal formed by the bacterium.

When susceptible insects—usually immature larvae—ingest Bt, the protein toxin is activated by alkaline conditions and enzyme activity in the insect's gut. Poisoned insects may die quickly from the activity of the toxin but usually die within one to three days from blood poisoning (bacterial septicemia).

Bt That Kills Caterpillars

Caterpillars are generally pests of greenhouse plants only at certain times of the year, but Bt can be used very effectively in controlling them. *Bacillus thuringiensis* var. *kurstaki* can be used to control cabbage looper, corn earworm, European corn borer, Florida fern caterpillar, salt marsh caterpillar, and pansy caterpillar. Common trade names for commercial products include Dipel, Javelin, Thruicide, and Bactospeine.

Products containing *Bacillus thuringiensis* var. *aizawai* are effective against several caterpillars that are only slightly susceptible to the *kurstaki* strain, such as the beet armyworm.

Bt That Kills Fungus Gnats

Bacillus thuringiensis serotype 14 (= *israelensis*) is effective in killing early-instar larvae of fungus gnats. This bacterium is sold under the commercial name Gnatrol for use in greenhouses. Repeated applications are often necessary since fungus gnats often have overlapping generations. Soil drenches of Gnatrol can be an effective means of fungus gnat control on crops that are kept at high moisture levels, such as plugs. Herb plant and vegetable transplant growers commonly use it where few standard pesticides are labeled.

Bt That Kills Beetles

Another Bt isolate is *Bacillus thuringiensis* var. *tenebrionis*, which is toxic to certain beetle larvae.

Ecogen Company has a Bt strain EG7673 that has genetically engineered strains of Bt that effectively kill leaf-feeding beetles. The product Raven has a label for use in agriculture, though its label for nursery and landscape use is still pending. This might have potential for greenhouse growers producing herbaceous perennials that are susceptible to feeding injury from leaf beetles.

Bt and Gene Therapy

Recent advances in biotechnology have allowed scientists to actually place Bt toxins within plants, creating what are called "transgenic plants." There are several methods for introducing genes into plants, including infecting plant cells with plasmids as vectors carrying the desired gene or shooting micro-scopic pellets coated with the gene directly into the cell ("gene gun" technology). Genes directing the production of Bt toxins can be incorpo-rated into plant-dwelling bacteria. When the altered bacteria grow in the inoculated plant, the Bt toxin is produced within the plant. This has been done with corn to protect the plants from corn borer. Although this appeared to be the ideal way to deal with pests susceptible to Bt, public senti-ments are tending away from this approach. Regardless, it would be years before this would be an economically feasible method for ornamental plants.

Effective Treatments

Each Bt controls only certain types of insects. It's essential to correctly ID your pests and determine whether Bt is suitable. Bt is generally most effec-tive on early, immature stages of susceptible insects. Since susceptible insects must consume Bt to be poisoned, treatments must be directed to the plant parts that the pests will eat. For target pests that bore into plants without feeding much on the surface or foliage, Bt is usually not very effective.

To maximize the effectiveness of Bt treatments, sprays should thoroughly cover all plant surfaces, including the undersides of foliage. Formulations of Bt that encapsulate the spores or toxins in a granular matrix or within killed cells of other bacteria provide protection from ultraviolet radiation, extend-ing the efficacy of the Bt.

Pest Management, May 2000.

Getting More from Beneficial Nematodes

Edwin Lewis

Beneficial (or entomopathogenic) nematodes have been available as a micro-bial insecticide for nearly twenty years. Growers and landscapers have used these beneficial pathogens against a variety of insect pests with varying degrees of success, some quite spectacular. For instance, the black vine weevil, one of the most serious pests in nursery crops, can be effectively controlled with nematodes. In fact, nematodes are one of a few effective treatments for the soil-inhabiting grub stage of this pest.

However, there are instances where nematodes should have provided acceptable control but didn't. Why are the results so unpredictable? The answer is often a lack of knowledge about how to store and apply nematodes.

It's impossible in a short article to address all of the situations growers may encounter. But if you're armed with accurate information about nematode biology, you can make educated decisions about when and where to use beneficial nematodes on a case-by-case basis—and maybe more importantly, when and where not to use them.

What Nematodes Do

When nematodes are formulated, packaged, and sold as microbial insecticides, it's easy to forget that they're actually living animals that have special requirements for staying alive and being able to infect an insect. Nematodes are sold when they're in the infective juvenile stage. This is the third of six life stages and is the only time nematodes can survive outside of an insect host. The infective juvenile nematode doesn't feed, mate, or develop outside of a host.

When you apply nematodes, you release the infective juvenile into the area you want to protect. The nematodes search out insect hosts and enter them through the mouth, anus, or spiracles. Once inside the host insects, the nematodes release bacteria, which they carry in their gut. The bacteria develop by destroying the insect tissue, killing the insect within two days. The nematodes develop by feeding on the bacteria. They reproduce for one to three generations within the host, and ten days later up to 500,000 infective juveniles emerge from the host ready to infect a new batch of host insects.

Nematode Success Tips

Here are a few rules of thumb when considering nematodes for insect pest management.

Product shelf life is limited. Check label recommendations, and don't use product that's too old—you may be applying dead nematodes.

Nematodes are soil organisms. This means they're effective against insects in protected habitats—soil insects, such as dark-winged fungus gnat larvae, and boring insects, such as dogwood borer. They don't tolerate dry conditions or direct sunlight. Attempts to use them against foliar insects have met with limited success. When treating soil in potted plants, water the crop before applying the nematodes. Watering after application can help, but be careful not to wash the nematodes out of the pot. To avoid problems with UV light in outdoor production areas, apply nematodes at the end of the day to give them time to get into the soil during low light levels. If you apply nematodes to landscapes, irrigate immediately after application.

The standard field application rate for nematodes is 1 billion nematodes per acre. When scaled to a per-pot basis, this works out to 20,000 nematodes for a twelve-inch pot. Manufacturer recommendations may differ from this figure, and those should be followed.

Nematodes available for sale won't be effective at soil temperatures cooler than 60°F. Make sure the soil is warm enough for the nematodes to be effective before you apply them.

All nematodes are not the same. Some species are effective against insects that live close to the soil surface, and others are equipped to infect insects deep in the soil. *Steinernema carpocapsae,* the most widely available nematode, is effective when used against surface-dwelling insects, such as black cutworm, but is ineffective against insects deep in the soil, such as white grubs. For insects deep in the soil, *Heterorhabditis bacteriophora* or *Steinernema riobravis* are effective. The nematode species in a product should be listed on the "active ingredient" portion of the label. If it isn't listed on the packaging label, contact the manufacturer and ask.

When using nematodes, monitoring pest populations is key to your success. Preventative treatments with nematodes don't work. The life expectancy of entomopathogenic nematodes in the field isn't well known, so apply them when hosts are present to ensure their effectiveness.

As with all IPM programs, record product performance and if you were satisfied with the product and the results. Keep track of application techniques that work and that don't work, and your effective use of entomopathogenic nematodes will improve from year to year.

Editor's note: For more on using beneficial nematodes, the author recommends the Insect–Parasitic Nematode Web site (www2.oardc.ohio-state.edu/nematodes). It's maintained by Dr. Parwinder Grewal (grewal.4@osu.edu), Department of Entomology, The Ohio State University.

Pest Control, April 1999.

Biocontrols for Poinsettia Diseases

Jae-soon Hwang and Mike Benson

Disease pathogens such as *Botrytis, Pythium,* and *Rhizoctonia* continue to plague growers. While excellent chemical controls exist and new ones are in the pipeline, researchers are still hard at work developing new ways to control diseases using biological controls. Biocontrols take advantage of environmentally safe microorganisms, predators, or parasites that are applied during crop production for disease and insect control. Several microbial biocontrol agents such as *Gliocladium, Trichoderma, Streptomyces, Burkholderia cepacia,* and binucleate rhizoctonia fungi have already been proven to work well against some disease organisms. some agents such as SoilGard (*Gliocladium*), Rootshield (*Trichoderma*), and Mycostop (*Streptomyces*) are gaining in use throughout the greenhouse industry.

Now the industry needs to develop the technology to extend the range of effectiveness of these and other biocontrols against multiple pathogens, as well as develop formulations that provide suitable shelf life, efficacy, and user-friendliness. The better these products work and the easier they are to use, the more growers will use them. And growers who adopt biocontrol agents as part of their overall IPM arsenal will enjoy the advantage of short reentry periods as well as a low worker exposure hazard. Plus, fungi that cause plant diseases don't develop resistance to biocontrol agents, so the long-term use and effectiveness of biocontrols isn't an issue.

The Approach at NCSU

Research in the Department of Plant Pathology at North Carolina State University, Raleigh, funded in part by a grant from the American Floral Endowment Inc., is aimed at integrating two biocontrol agents to manage three poinsettia diseases: *Pythium* root rot, *Rhizoctonia* stem and root rot, and *Botrytis* blight.

In our preliminary experiments, a series of applications of a strain of the bacterium *Burkholderia* (*Pseudomonas*) *cepacia* called 5.5B and a nonpathogenic strain of binucleate *Rhizoctonia* fungus (BNR) on poinsettias has shown promise for control of *Rhizoctonia* stem rot during propagation and rhizoctonia root rot after transplanting. An ever-present pathogen in most greenhouses, *Rhizoctonia* can cause significant losses during propagation and after transplanting rooted cuttings.

Learning How Biocontrols Work

Understanding how these biocontrol agents prevent *Rhizoctonia* infection is the key to making the technology more effective and reliable for use in the floriculture industry.

Previous work in our lab has shown that strain 5.5B of *Burkholderia cepacia* produces an antifungal compound known as pyrrolnitrin. To learn more about the role of pyrrolnitrin in protecting poinsettia cuttings from *Rhizoctonia* stem rot, we selected nine rifampicin-resistant mutant strains of 5.5B (this type of mutant can be easily tracked in the laboratory). Then we tested these nine mutant strains in a greenhouse experiment to see if they would control stem rot on poinsettia cuttings.

First, the original strain 5.5B and each mutant strain were multiplied and mixed with distilled water to a specific concentration. Next, we soaked rooting cubes with cell suspensions of the individual strains. Then we stuck poinsettia cuttings into the cubes and placed a small amount of *Rhizoctonia* on the cube about a half-inch away from the cutting to inoculate it.

We found that the effectiveness of biocontrol varied depending on the particular mutant we tested. For instance, strain 21-2 showed the same level of disease control as the original "wild-type" strain 5.5B, while strain 13-1 showed no biocontrol activity toward stem rot at all. The rest of the strains varied in their ability to control stem rot.

Correlating Our Findings

Back in the lab, we examined the pyrrolnitrin production of each of these nine mutants using a technology called "thin layer chromatography" (TLC). Cultures of each mutant were extracted and were treated with a special chemical on TLC plates. The original strain and strain 21-2 produced the same distinct purple spot on the TLC plate, indicating that both produce pyrrolnitrin at the same concentration. However, only a very faint spot was detected when we analyzed the ineffective strain 13-1.

This leads us to believe that how well a *Burkholderia cepacia* strain controls *Rhizoctonia* may be related to the amount of pyrrolnitrin it produces. But we need to do more testing of the precise amounts of pyrrolnitrin production by each mutant using a more sensitive test method to confirm the role of pyrrolnitrin in disease control.

What about Multiple Applications?

With most plant diseases, it can be difficult to get good, dependable, season-long disease control using a single application of a fungicide or biocontrol agent. To extend the period of crop protection, techniques must be developed that use additional applications of the same biocontrol agent, combine chemicals with biocontrol agents, or combine the application of two or more biocontrols.

BNR, which is a *Rhizoctonia* fungus, has been shown to be an effective biocontrol for poinsettia stem rot. BNR is similar to the type of *Rhizoctonia* that infects your crops except for one important characteristic: BNR doesn't cause disease in plants. We've tried combining BNR with 5.5B for double-barreled biocontrol.

To test the effectiveness of combined applications, we treated poinsettia cuttings in rooting cubes with 5.5B to control *Rhizoctonia* stem rot. At transplanting, we amended the media with a 0.6% formulation of BNR to see how well it would protect the poinsettias during the remainder of the crop production cycle.

Treating the rooting cubes with 5.5B reduced the severity of *Rhizoctonia* stem rot by 57% compared with the untreated controls. And incorporating BNR into transplant media effectively controlled *Rhizoctonia* for up to eight weeks after transplanting. The severity of *Rhizoctonia* in the BNR-amended mix was 67% lower than in the infested controls. This sequential combo of biocontrols, one a fungus and one a bacterium, worked better than using the bacterium (5.5B) on rooting cubes and then repeating the same treatment at transplant.

Further experiments will look for the changes in the population of the biocontrol agents to determine if follow-up applications will be required after the initial applications.

August 1998.

Beneficial Viruses—New Allies in the Fight against Bacterial Diseases

B. K. Harbaugh, J. B. Jones, J. E. Flaherty, and L. E. Jackson

When the diagnosis comes back from the plant clinic that a bacterium is responsible for the disease, the first thoughts of most growers producing floricultural crops would probably be that the only good bacteria are dead ones and that they have a serious problem. Unfortunately, many bacterial pathogens on floricultural crops can be very difficult and costly to control. Current control measures are often directed at preventing disease, as few control options are available to growers. Preventative strategies include using culture-indexed plants, destroying infected plants, disinfecting equipment and growing structures, and growing other crops that aren't susceptible to the bacterial pathogen.

Our research at the Gulf Coast Research and Educational Center, the University of Florida, and AgriPhi Inc., which is supported by the American Floral Endowment, is a novel approach utilizing bacteriophages (also known as phages) as biological control agents for the prevention and control of bacterial diseases of ornamentals. While we often cringe at the thought of viruses, bacteriophages are viruses that kill only bacteria (fig. 1). These beneficial viruses are specific for target bacteria and are nontoxic to workers and nontargeted bacteria. Thus, this research was initiated to develop an environmentally safe and effective alternative for prevention and control of bacterial pathogens by utilizing bacteriophages.

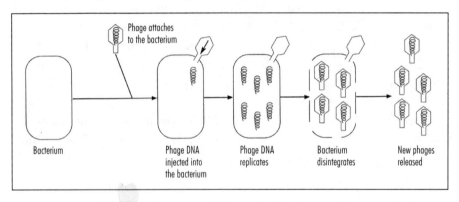

Figure 1. The life cycle of bacteriophages (also known as phages), which are viruses that kill bacteria.

The idea of using bacteriophages for biological control of bacterial plant pathogens goes back to 1926. However, bacteriophages were often applied as individual isolates, and resistant strains of bacteria developed rapidly, as they do with chemical bactericides or antibiotics. To prevent this, our approach encompasses a method developed and patented by Dr. Lee Jackson that uses a mixture of h-mutant bacteriophages that have been selected because they have a broad host range.

The process of developing an effective mixture of bacteriophages begins with a search for wild-type bacteriophages in various environments including soil, water, and sewage sludge. After isolation, wild-type bacteriophages are tested against many different selections of bacteria. When resistant strains of bacteria develop, high populations of wild-type bacteriophages are introduced. Due to the spontaneous mutations that occur in bacteriophages, it's possible to select for mutant bacteriophages that can kill the formerly resistant strain of bacteria (fig. 2).

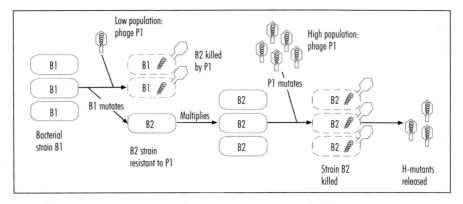

Figure 2. Development cycle of h-mutant phages. H-mutant phages originate from spontaneous mutations of phages. They have a broader host range than wild-type phages, and in the example above would be able to kill both B1 and B2 bacterial strains.

Select mutant bacteriophages are able to kill both the original parental type and resistant bacteria. We call them h-mutant, as they were selected mutants from wild-type bacteriophages and they have an expanded host range compared to the original wild-type bacteriophage. As further insurance against bacteria developing resistance, three to eight h-mutant bacteriophages are combined in a mixture when applied for control purposes.

Geranium Blight: A Model System

Perhaps the most difficult problem to overcome in geranium production, and certainly the most serious disease of geraniums, is bacterial blight caused by *Xanthomonas campestris* pv. *pelargonii* (Xcp). Geranium growers fear bacterial blight, also referred to as bacterial stem rot, bacterial leaf spot, or bacterial wilt, because it's extremely difficult to control. We're currently investigating the use of bacteriophages for the biological control of geranium bacterial blight. These studies serve as a model system for using bacteriophages on potted crops grown in the greenhouse or field.

We've successfully isolated bacteriophages for Xcp from sewage sludge and soil samples taken from pots and ground beds that contained geraniums with bacterial blight. Bacteriophages have been isolated from samples from California, Florida, Utah, and Minnesota. After the selection of these bacteriophages for increased virulence, seventeen bacteriophage isolates were screened against twenty-one Xcp strains that were collected from samples of geraniums from around the world. Five bacteriophages produced virulent reactions in at least seventeen Xcp strains. A mixture of five of the best bacteriophages was developed and again tested against these twenty-one Xcp strains. All twenty-one Xcp strains were sensitive to the mixture of bacterio-phages. Thus, we've produced a mixture of bacteriophages with a broad host range that has allowed us to initiate biological control tests in the production environment.

In the Greenhouse

A number of greenhouse experiments have been completed to demonstrate the biological control potential of the bacteriophage mixture developed for *Xanthomonas* blight of geranium. For example, 'Maverick Red' geranium plants were inoculated with Xcp, then placed in the center of a group of eight non-inoculated plants growing in 4 1/2-in. pots. The bacteriophage mixture was applied daily or twice weekly, and the development and spread of Xcp were compared to plants treated with Phyton-27 (a copper-based bactericide labeled for Xcp control on geraniums), applied at two-week intervals. Table 1 presents some results from this test and illustrates the success we're achieving with the bacteriophage mixture.

Table 1. Effectiveness of Bacteriophages on *Xanthomonas* of geraniums					
Disease severity	Noninoculated control	Inoculated control	Phyton-27 (1.5 oz./10 gal.)	Phage daily	Phage biweekly
Plants with lesions	0.5 a	4.3 c	4.3 c	1.5 b	4.0 c
Lesions/eight plants	0.5 a	7.2 c	9.2 c	1.8 b	10.0 c

There was an almost threefold reduction in the number of plants that became infected and a fivefold reduction in the number of lesions for plants treated with bacteriophage daily, compared to plants in the inoculated control or copper treatments. We've found that this biological control method has consistently given equal or better control than the copper-based bactericide. While the bacteriophage mixture didn't eradicate Xcp, and Xcp did spread slightly, it's important to realize that conditions favored the spread of Xcp (i.e., inoculated plants were left in the middle of test plants for eight days, and overhead irrigation was used to help spread the bacteria).

We were concerned that because the bacteriophage mixture didn't completely eliminate the spread of Xcp, a resistant strain of Xcp may have developed. We reisolated Xcp from diseased plants treated with the bacteriophage; these Xcp isolates were susceptible to the phage mixture. Thus, we believe that resistance wasn't the reason for the lack of control.

Because we didn't achieve comparative control with twice weekly sprays compared to a daily application of bacteriophages, we've initiated tests to look at application methods, phage concentrations, and timing of applications. These preliminary results also indicate that we need to improve the longevity of the bacteriophages on leaf surfaces or in the production environment so that fewer applications are necessary.

Future Plans

While these initial results have shown promise, we realize that much more research is needed before the full potential of this new control strategy can be realized. We're still seeking and analyzing plant and soil samples from growers around the U.S. and the world.

We're also initiating studies on *Erwinia* soft rot of poinsettias (incited by *Erwinia carotovora* [Ec]) as a model system for control of bacterial pathogens on crops in a propagation (mist) house. It's important to find many strains of bacterial pathogens and bacteriophages to ensure that a viable mixture of bacteriophages can be developed for commercial use.

We would welcome samples of drainage water, geraniums with Xcp or poinsettias with Ec, or soil from under benches or from pots with diseased plants. (Yes, you will remain anonymous.)

As more bacteriophage strains are isolated, we'll continue our work to demonstrate the possibilities of using appropriate mixtures of h-mutant phages for biological control of bacterial diseases. We anticipate that information gained from these model systems can be applied to other floricultural crops including cut flowers, flowering pot plants, and bedding plants.

October 1998.

Chapter 3
Chemical Controls

Explaining Chemical Formulations
Ronald D. Oetting and Denise L. Olson

The active ingredients of most inorganic pesticides are pure chemicals that may be either liquid or solid and may or may not be water soluble. The active ingredients must be formulated to provide a pesticide that meets the requirements of you, the end user. This means it must have an acceptable shelf life, be miscible (able to be mixed in water), and be easily handled and applied to the target. This is accomplished by mixing other ingredients with the active ingredients—the result is the formulation.

The name of a pesticide is usually followed by a number or numbers that provide information on the concentration of the active ingredients in the formulation and a letter or letters that designate what type of formulation. For example, in Diazinon 2E, the "2" represents the concentration of the active ingredient expressed as pounds of active ingredients per gallon. The "E" represents the formulation, an emulsifiable concentrate.

The material added to the active ingredient is called a diluent. Diluents are used to extend the active ingredient and are usually considered inert. Actually, they function chemically, physically, or physiologically to influence the effectiveness of the active ingredient. Some diluents are carriers used for their absorptive capacity. These carriers, such as talc and clay, absorb the liquid active ingredient to produce granular and powder formulations. Some diluents are emulsifiers, which allow active ingredients that aren't miscible with water to remain in suspension while applied.

Liquid Formulations
Emulsifiable concentrates are the most common liquid pesticide formulations. Emulsifiable concentrates aren't miscible in water and must have an emulsifier added to keep the active ingredient in suspension. Dursban 2E is one example.

Another common liquid formulation is the water soluble. This formulation is at least partially miscible in water. Vydate 2L is one example. The "2" represents the concentration of active ingredients in pounds per gallon, and the "L" represents a liquid formulation. Some active ingredients have been encapsulated in microcapsules (M), providing a slow release through the membrane of the capsule, such as Knox-Out 2FM.

Solid Formulations

The primary solid formulation is wettable powder (WP), for example, Thiodan 50WP. In this case, the number in the trade name represents the percent active ingredients. Therefore, 50 percent of Thiodan WP is active ingredient; the remainder is the inert carrier of the active ingredients. These materials must also have wetting agents or be agitated to keep the solid particles in suspension in the spray tank so they don't settle to the bottom.

Because wettable powders are dusty, absorb moisture, and aren't easy to measure and handle, many manufacturers have changed their WP formulations to flowable. A flowable is a wettable powder ground into finer particles and formulated into a thick, milky solution. The fine powder allows greater distribution on the target, and the flowable formulation eliminates the dusty powder. The solids in flowables can settle and need to be agitated before using.

Some wettable powders have been formulated into granules that dissolve in water, releasing the particles of active ingredients. Some have been premeasured in soluble bags that can be dropped into spray tanks. Some solid forms of active ingredients are water soluble, for example, Orthene 75S.

Dusts and Granules

Not all pesticides are sprayable. Dusts (D) and granules (G) are applied directly out of the bag to the plant or soil. Dusts are very similar to wettable powders but have a much lower concentration of active ingredients. Granules are also low-concentration formulations that are applied directly to potting media. They're usually systemic.

Pesticides are available in many different types of formulations. Many formulations are similar in character, but slight alterations make them different in terms of shelf life, water miscibility, and ease of handling and application. The following is a list of letters that designate various pesticide formulations.

B = bait	EC = emulsifiable concentrate
D = dust	EL = emulsifiable liquid
DG = dispersible granules	F = flowable (aquaflow)
E = emulsifiable	FC = flowable concentrate

FM = flowable microencapsulated
G = granules
L = liquid
LC = liquid concentrate
M = microencapsulated
S = sprayable powder
SC = spray concentrate
SP = soluble powder

W = wettable powder
WBC = water-based concentrate
WDL = water-dispersible liquid
WM = water miscible
WP = wettable powder
WSB = water-soluble bags
WSP = water-soluble packets

Some manufacturers have added letters at the end of trade names. The letters don't have anything to do with formulation but indicate what type of usage this active ingredient is registered for:

CS = commercial applicators
GH = greenhouse
GC = golf course
O = ornamentals

T&O = turf and ornamentals
TT&O = turf, tree, and
    ornamentals

Pest Control, April 1997.

Beyond the Label: Maximizing Pesticide Performance

Jim Willmott

Pesticides are essential pest management tools. Effective results, however, require proper use. Too often, failures are attributed to products, while application mistakes actually caused the problem.

Here are some general ideas for improving pesticide performance, along with specific tips for success with twenty-five fungicides, insecticides, and miticides that were reported in a recent survey (Hudson, et al., 1996, Jones, et al., 1996) as the products most frequently used by the U.S. greenhouse and nursery industry (imidacloprid, not included in the survey, was added to our list due to its widespread use). Pesticides are listed in alphabetical order by their common names—not in order of reported use.

Before You Spray

The key to pesticide success is to keep in mind that chemicals should always be used as part of an integrated pest management (IPM) program. Neglecting the essentials—pest monitoring, exclusion, sanitation, biological control, and good cultural practices—reduces pesticide results and increases the probability of pest resistance. With that said, here are nine more factors that are critical to effective pest control.

1. Make the correct diagnosis. Don't jump to conclusions! Misidentification often results in further crop damage. For example, you might assume mites are damaging ivy geraniums, when they actually have edema, a physiological problem. Miticides won't solve the cultural and environmental problems that cause edema. If you're not sure, get help.

2. Detect the problem early. This requires formalized monitoring and knowledge of key crops and pests. Early detection reduces crop damage and pesticide use.

3. Select the best pesticide. This depends on many factors, but each pesticide has strengths and weaknesses. Performance varies with factors such as environmental conditions and stage of pest or crop development. Marathon, for example, is an excellent systemic whitefly insecticide, but it's not the best choice for young poinsettia cuttings with limited root development because roots absorb it. Remember: Don't rely on one product. Rotate between chemical classes to reduce risk of pest resistance. For insects and mites, rotate after two or more applications of the same product. And rotate your fungicides, too. While it's not stressed as often as for insecticides, it's equally important.

4. Read the product label thoroughly. Many grower mistakes could be avoided by following label instructions. For example, a 1996 survey of greenhouse pesticide use indicated that 20% of growers applied Subdue as a foliar spray. Yet, with the exception of azaleas, Subdue is labeled exclusively for media drenches for soil-borne pathogens on greenhouse crops. Another common and potentially serious mistake is treating vegetable transplants and herbs with pesticides that haven't been tested or approved for them.

5. Select the appropriate application equipment. Effective equipment results in good coverage of plant parts infested or infected by the target pest. Hydraulic sprayers, mist blowers, electrostatic sprayers, injectors, foggers, aerosol cans, and smokes each have their place and are effective for the right pest at the right time.

6. When possible, start small. Avoid treating large areas all at once—especially when using new products and tank mixes or when treating new and unfamiliar crops.

7. Understand your water. Most growers recognize that water quality is important to crop nutrition, but few are aware that it sometimes relates to pesticide efficacy. High water pH and alkalinity, or hardness from calcium, magnesium, or other ions, can result in reduced pesticide efficacy. For example, insecticidal soaps precipitate when mixed in water with high calcium and magnesium. Also, some pesticides, including Dycarb, are quickly hydrolyzed in alkaline water. Check your water and adjust pH and alkalinity with appropriate adjuvants. Always mix and apply pesticide solutions promptly to reduce the chance of breakdown.

8. Purchase appropriate quantities and store them in a cool, dry, well-ventilated area. Most pesticides have a shelf life of several years. However, improper storage conditions and long time periods increase the potential for failure because of mixing problems or breakdown of active ingredients. Be sure to read and follow label directions for proper storage, and don't buy more than you need for one year.

9. Keep accurate records of applications and results. Record all information including pesticide rates, application equipment, crops, environmental conditions, time of day, etc. Don't forget to evaluate application results. Document results along with application information. Good records are essential for planning future pest management strategies.

Inclusion or omission of products is based on the results of the survey and does not constitute endorsement by Rutgers University Cooperative Extension. Pesticide applicators are responsible for ensuring that use complies with regulations and the product label.

Tips for Specific Products: Insecticides and Miticides

Abamectin
- **Trade name:** Avid 0.15 EC
- **Pesticide class:** macrocyclic lactone
- **Key pests:** spider mites and leafminers
- **Label crops:** most greenhouse and field-grown ornamentals
- **REI:** twelve hours

When applied to upper leaf surfaces, Avid penetrates into tissue and kills mites that inhabit and feed on lower leaf surfaces. While it's chemically unrelated to other miticides and insecticides, resistance to Avid has been reported. To reduce the risk, Dr. David Smitley, Michigan State University, recommends rotating miticides so that each one is used for four weeks before switching to a new one. Use three or more products in the rotation. Possibilities include Kelthane, Pentac, Sanmite, M-Pede (soap), Sunspray (horticultural oil), Attain, Talstar, and Mavrik. Don't apply Avid to ferns and Shasta daisies.

Acephate

- **Trade names:** Orthene 75SP, PT 1300, Orthene, PT 1300 DS, Orthene, and Pinpoint 15G
- **Chemical class:** organophosphate
- **Key pests:** whiteflies, thrips, aphids, scales, mealybugs, and leafminers
- **Labeled crops:** varies with formulation (see labels)
- **REI:** twenty-four hours

Acephate is a systemic insecticide that penetrates plant tissue and moves upward and outward in plants. Good activity is reported on many sucking pests including aphids and whiteflies. University and industry research has demonstrated enhanced activity against aphids, whiteflies, and thrips when acephate is tank mixed with various pyrethroid insecticides. Orthene SP and Tame are labeled for tank mixing. The combination is synergistic, giving increased control over either product used alone. Orthene activity is also enhanced when combined with insecticidal soap, but the risk of plant injury increases. Aerosol formulations, for use only in greenhouses, can also be "space mixed"—released into the greenhouse at the same time. PT 1300 Orthene can be space mixed with PT 1800 Attain or PT 1200 Resmethrin. Regardless of application technique, good coverage ensures optimal results. When using aerosol formulations, use fans to create horizontal air circulation.

Bacillus thuringiensis *(Bt)*

- **Trade names:** Gnatrol
- **Chemical class:** microbial/biological
- **Key pests:** fungus gnat larvae
- **Labeled crops:** most greenhouse ornamentals
- **REI:** twelve hours

Gnatrol is commonly used to control fungus gnat larvae. It contains a highly effective spore-forming strain of Bt that produces toxic crystals that are lethal when consumed by fungus gnat larvae. Populations are difficult to control once larvae penetrate into roots and crowns since they are less likely to ingest the Bt. Bt doesn't control adults. This product should be applied as a drench toward the end of an irrigation period to avoid dilution or leaching. Avoid applications in combination with fertilizers or fungicides containing copper or chlorine since they may neutralize the active ingredient. Applications must be repeated every few days until infestations are controlled.

Bifenthrin
- **Trade names:** Talstar T&O 10 WP and F, PT 1800 Attain
- **Chemical class:** pyrethroid
- **Key pests:** aphids, mealybugs, whiteflies, fungus gnats (adults only), scales, caterpillars, and spider mites
- **Labeled crops:** many greenhouse and outdoor ornamentals
- **REI:** twelve hours

Bifenthrin is a pyrethroid insecticide with miticidal properties. Since it's not systemic, good spray coverage is key to successful pest control with Talstar. Applications must be directed to cover lower leaf surfaces for whiteflies and mites. Talstar activity may be increased by tank mixing with organophosphates. Many growers tank mix with Azatin, an insect growth regulator (IGR), for control of both adult and larval stages of fungus gnats. PT 1800 is an aerosol formulation of bifenthrin. Apply it to dry leaf surfaces when temperatures are between 65° and 80°F. It may be applied alone or in combination, as a space mix, with PT 1300 Orthene or PT 1200 Resmethrin.

Chlorpyrifos
- **Trade names:** PT 1325 ME DuraGuard
- **Chemical class:** organophosphate
- **Key pests:** aphids, caterpillars, fungus gnat larvae, leafminers, mealybugs, scales, thrips, and whiteflies
- **Labeled crops:** many greenhouse and field-grown ornamentals
- **REI:** twelve hours

DuraGuard is a relatively new formulation of chlorpyrifos that is microencapsulated to improve residual activity and plant safety. As a contact insecticide, good coverage is necessary. Be sure to shake the container prior to adding concentrate to the spray tank. Maintain agitation of spray solution to prevent settling in tank and uneven application. DuraGuard provides excellent control of fungus gnat larvae.

Cyfluthrin

- **Trade name and formulation:** Decathlon 20 WP
- **Pesticide class:** pyrethroid
- **Key pests controlled:** aphids, caterpillars, fungus gnat adults, mealybugs, and thrips
- **Label crops:** many greenhouse and field-grown ornamentals
- **REI:** twelve hours

Decathlon is a pyrethroid insecticide with contact activity, so complete spray coverage is necessary. Maintain agitation in spray tank. Tank mixed combinations with organophosphates such as Orthene may improve control of some pests, including silver leaf whitefly. Very low use rate of 1 to 2 oz. per 100 gal. requires careful measuring.

Diazinon

- **Trade names:** Knox-Out 2FM, Knox-Out PT1500R
- **Chemical class:** organophosphate
- **Key pests:** aphids, caterpillars, fungus gnat larvae, leafminers, mealybugs, scales, spider mites, thrips, and whiteflies
- **Labeled crops:** many greenhouse and field-grown ornamentals (see label)
- **REI:** twelve hours

Diazinon is a contact insecticide with no systemic properties. Thorough spray coverage is necessary. Be sure to cover lower leaf surfaces. Microencapsulated formulation gives Knox-Out improved residual activity and plant safety.

Dienochlor

- **Trade name:** Pentac Aquaflow 4F, Pentac 50 WP
- **Pesticide class:** chlorinated hydrocarbon
- **Key pests controlled:** mites and whiteflies
- **Label crops:** many greenhouse ornamentals
- **REI:** twenty-four hours

Pentac is a contact miticide that also has activity against whitefly eggs and nymphs. As a miticide, its results are slower than for other products. Thorough spray coverage is essential. Ensure adequate mixing and agitation in spray tank to prevent settling and uneven application. For high populations under warm conditions, repeat application should be made in seven days. Combination with a pyrethroid with miticidal activity will improve knockdown and reduce the risk of resistant mites. Many growers tank mix Pentac with Mavrik.

Endosulfan

- **Trade name:** Thiodan 50 WP & 33 EC, Fluex Thiodan Smoke
- **Pesticide class:** chlorinated hydrocarbon
- **Key pests controlled:** aphids, cyclamen mites, and whiteflies
- **Label crops:** many greenhouse and field-grown ornamentals and tomatoes (smoke formulation for greenhouse only)
- **REI:** twenty-four hours

Thiodan is one of the few remaining chlorinated hydrocarbon pesticides. As such, it's useful in insecticide rotations that often contain organophosphates and pyrethroids. While not indicated on the label, some growers report successful thrips control. All endosulfan formulations are highly toxic: Follow all labeled safety precautions when mixing and applying. Sprayable formulations require good coverage. Applications may injure open blooms. Thiodan is also available as a smoke formulation in Fluex Thiodan. It produces extremely small particles that disperse uniformly throughout enclosed greenhouses before settling on plant surfaces. Be sure that temperatures are above 70°F and below 90°F during application. Also, leaf surfaces must be dry. While some smokes require only that EPA ventilation criteria be met, Thiodan smoke has a twenty-four-hour REI.

Fenpropathrin

- **Trade name:** Tame 2.4 EC
- **Pesticide class:** pyrethroid
- **Key pests controlled:** aphids, whiteflies, mealybugs, and spider mites
- **Label crops:** anthurium, chamomile, bedding plants, chrysanthemum, columbine, foliage plants, geranium, gladiola, impatiens, lily, marigold, poinsettia, and snapdragon grown in greenhouses, lath houses, and shade houses only
- **REI:** twenty-four hours

Tame is a contact insecticide. It's labeled for tank mixing with Orthene. Follow label precautions to avoid phytotoxicity. Rotate with pesticides in other classes after several applications to avoid resistance. For easier measuring of small quantities, convert ounces to milliliters and measure with a graduated cylinder. One ounce equals 29.6 milliliters.

Fluvalinate

- **Trade name:** Mavrik Aquaflow 2F
- **Pesticide class:** pyrethroid
- **Key pests controlled:** aphids, caterpillars, mealybugs, spider mites, thrips, and whiteflies
- **Label crops:** many greenhouse and field-grown ornamentals
- **REI:** twelve hours

Mavrik is a contact insecticide/miticide requiring good coverage especially for whiteflies and spider mites that inhabit lower leaf surfaces. Adjust spray water pH to between 5.0 and 7.0. Mavrik may be tank mixed. Many growers combine with Pentac for improved mite control, and with Enstar II for improved control of immature aphids, fungus gnats, whiteflies, and scales.

Horticultural oil

- **Trade name:** Sunspray Ultrafine
- **Pesticide class:** horticultural oil
- **Key pests controlled:** aphids, fungus gnat adults, leafminers, mealybugs, scales, spider mites, thrips, and whiteflies
- **Label crops:** most greenhouse and field-grown ornamentals, vegetables, and herbs (see label)
- **REI:** twelve hours

Spray must completely cover target pests. Avoid applications during periods of high temperatures and humidity. Don't mix with fungicides or any sulfur-containing pesticide. Agitation of spray solution must be maintained. Since there is no residual control, repeat treatments weekly until pest populations are under control. Tank mixes with insecticides often increase effectiveness, but potential for plant injury also increases.

Imidacloprid

- **Trade name:** Marathon 1G & 60 WP
- **Pesticide class:** chloronicotinyl
- **Key pests controlled:** aphids, mealybugs, scales, thrips, and whiteflies
- **Label crops:** many greenhouse and field-grown ornamentals
- **REI:** twelve hours

Over the last few growing seasons, Marathon has established itself as a reliable systemic insecticide for controlling a wide range of greenhouse insect pests. The product is absorbed through roots and translocates to foliage.

Marathon formulations are labeled for soil treatment only: Do not apply WP formulation to foliage! Successful application requires well-developed, healthy root systems. Applications prior to rooting are prone to leaching. Marathon is long lasting, but relatively slow to act. Apply early in the crop cycle to young plants under conditions that favor water uptake. Avoid excessive irrigation and leaching for ten days following application. When applying WP formulation through irrigation systems, be sure to thoroughly mix product in stock tank and maintain agitation. Also, ensure that systems are calibrated and provide uniform distribution to crops.

Insect growth regulators (IGRs)
- **Trade names:** Adept 25 WP, Azatin 3 EC, Citation 75 WP, Dimilin 25 WP, Enstar II, Neemazad 4.5 EC, Precision 25 WP, PT 2100 Preclude

Survey results indicated that IGRs were used frequently by growers, but it didn't specify products. IGRs are effective in controlling the immature stages of several key pests, including thrips, whiteflies, aphids, fungus gnats, shore flies, and leafminers. Most are labeled exclusively for ornamental greenhouse crops, but the neem-based products Azatin and Neemazad may be used on many vegetables. Check labels for specific pests and crops and application techniques.

Insecticidal soap
- **Trade name:** M-Pede, Olympic Insecticidal Soap
- **Pesticide class:** potassium salts of fatty acids
- **Key pests controlled:** aphids, caterpillars, fungus gnat adults, mealybugs, mites, scales, thrips, and whiteflies
- **Label crops:** many greenhouse and field-grown ornamentals, vegetables, and herbs (see label)
- **REI:** twelve hours

Soaps work by contact—they have no residual activity, so be sure spray coverage is thorough. Water containing high levels of calcium, magnesium, iron, and other metallic ions that are associated with hard water will reduce effectiveness. Conditioning adjuvants can be added if needed. Applications are most effective when made during conditions that favor slow drying. Evening, early morning, and cloudy periods are ideal. Applications under high light, temperature, and humidity increase the chances for crop injury. Tank mixes with other insecticides often increase activity, but the risk of crop injury also increases.

Fungicides

Chlorothalonil

- **Trade name:** Daconil 2787 75 WP & 4F, Exotherm Termil 20% Smoke
- **Pesticide class:** carbamate
- **Key diseases or pathogens controlled:** *Botrytis,* black spot of rose, leaf spots including *Alternaria* and *Septoria,* powdery mildews, and rusts (see label for complete list)
- **Labeled crops:** many greenhouse and field-grown ornamentals (see label). Exotherm Termil is for greenhouse use only.
- **REI:** forty-eight hours

Chlorothalonil is a protectant contact fungicide that requires complete coverage of plant tissue. Maintain protection of new growth by applying at the recommended label intervals. Wet sprays with hydraulic sprayers work best with Daconil formulations. Mix well and maintain agitation. Exotherm Termil is a smoke formulation. Apply to dry plants when temperatures are between 70° and 90°F. Before igniting, be sure all vents are closed and cans are spaced properly and are set on fireproof surfaces. Start igniting cans at the point farthest from the door.

Copper sulfate pentahydrate

- **Trade name:** Phyton-27 5.5 EC
- **Pesticide class:** fixed copper
- **Key diseases or pathogens controlled:** *Alternaria,* black spot of rose, *Botrytis,* powdery mildew, leaf spots caused by *Pseudomonas,* and *Xanthomonas*
- **Labeled crops:** many greenhouse and field-grown ornamentals (see label)
- **REI:** twenty-four hours

Phyton-27 may be applied as a spray, fog, or drench. It has good plant safety and is used on tender plant tissue, including cuttings and flowers. Use of low-volume equipment is effective on *Botrytis,* but not on powdery mildew or *Xanthomonas.*

Etridiazole

- **Trade name:** Terrazole 35WP, Truban 30W, 25EC, 5G
- **Pesticide class:** triadiazole
- **Key diseases or pathogens controlled:** root and crown and foliage diseases caused by *Phytophthora* and *Pythium*

- **Labeled crops:** many greenhouse and outdoor ornamentals (see label)
- **REI:** twelve hours

Fungicide must be drenched into media to protect roots and crowns. For bedding plants in flats, mix recommended rate into 100 gal. of water and apply to 800 sq. ft. (approximately 1 qt. per flat). For potted plants, 100 gal. should cover about 400 sq. ft. (8 oz. per 6-in. pot). Water after application with about 50% of drench volume. Etridiazole and thiophanate-methyl are combined in Banrot for broad-spectrum control of root and crown rots. However, the highest label rates for Banrot contain less etridiazole than the suggested curative rate on Terrazole or Truban labels. If you have active *Pythium* or *Phytophthora,* the higher rate is necessary. Keep in mind, strict sanitary practices combined with judicious irrigation will go a long way in minimizing root rot problems. Don't rely exclusively on fungicides!

Fosetyl-aluminum

- **Trade name:** Aliette 80WP & 80WDG
- **Pesticide class:** organic phosphate
- **Key diseases or pathogens controlled:** *Pythium* and *Phytophthora* rots, downy mildew of rose
- **Labeled crops:** many greenhouse and outdoor ornamentals (see label)
- **REI:** twelve hours

Aliette is a systemic fungicide that, unlike most others, moves downward as well as upward in plants. Drench application is labeled for many crops. Foliar applications are specifically labeled for bedding plants. May be combined with Chipco 26019 for broader-spectrum pathogen control. Aliette WDG isn't compatible with Daconil 2787. To avoid copper phyto-toxicity, don't apply Aliette and copper-based products at less than seven-day intervals. Also, don't mix with any sticking, spreading, or wetting agent. Minimize root rot problems with good sanitation and watering practices.

Iprodione

- **Trade name:** Chipco 26019 50WP
- **Pesticide class:** dicarboximide
- **Key diseases or pathogens controlled:** *Alternaria, Botrytis,* and *Rhizoctonia*
- **Labeled crops:** many greenhouse and outdoor ornamentals including bedding and foliage plants (see label)
- **REI:** twelve hours

Contact fungicide with localized systemic activity. Apply as a spray, drench, or dip. Labeled for application through high-volume equipment. Foliar sprays give excellent control of *Botrytis,* but rotate fungicide classes to avoid resistance. Don't use exclusively in rotation with Ornalin or Curalan since active ingredients are in the same chemical class. Drench applications are effective on *Rhizoctonia,* but don't treat impatiens or pothos. Growers who apply through injectors and drip irrigation must be careful to maintain agitation of stock solution and ensure uniform distribution to crops.

Mancozeb

- **Trade name:** Protect T/O 80 WP, Dithane T/O 75DF, Dithane T/O 4F, Fore 80WP
- **Pesticide class:** carbamate
- **Key diseases or pathogens controlled:** *Alternaria,* black spot of rose, *Botrytis,* downy mildews, flower blights, and many leaf spots and rusts
- **Labeled crops:** many greenhouse and outdoor flowering pot crops, bedding plants, foliage plants, and perennials. Crops vary with product (see labels).
- **REI:** twenty-four hours

Protectant contact fungicide with no systemic activity. Ensure complete coverage of plant tissue. During conditions conducive to disease, maintain applications at seven- to ten-day intervals to ensure adequate protection of new growth. Use a spreader sticker for hard-to-wet foliage. Maintain agitation of solutions to prevent settling and uneven applications.

Metalaxyl/mefanoxam

- **Trade name:** Subdue Maxx, 2E & 2G
- **Pesticide class:** acylalanine
- **Key diseases or pathogens controlled:** *Pythium, Phytophthora,* and azalea petal blight
- **Labeled crops:** many greenhouse and field grown ornamentals including bedding, foliage, and flowering pot crops
- **REI:** twelve hours, except for Subdue Maxx, which has a zero-hour restricted-entry interval

Subdue is a systemic fungicide that is selective for the water mold fungi *Pythium* and *Phytophthora.* It should be applied to growing media, not as foliar spray, except for petal blight on azaleas. Applications should be made to media either through sprays or through irrigation systems, including drip types. Spray applications should be followed by a minimum of a half-inch of water. When applying through injectors, maintain agitation in stock solution and ensure accurate calibration and uniform distribution. Apply about 1 pt. of solution per sq. ft. to containers up to four inches deep. Larger containers need 1½ to 2 pt. per sq. ft. Subdue has good residual action, so application intervals are relatively long. Depending on the crop, apply every one to two months. To minimize the chances of phytotoxicity, measure carefully and don't exceed recommended rates or application intervals. Subdue Maxx has lower application rates than Subdue 2E.

PCNB

- **Trade name:** Terraclor 75WP, Terraclor 4F, PCNB 75WP
- **Pesticide class:** chlorinated hydrocarbon
- **Key diseases or pathogens controlled:** *Rhizoctonia, Sclerotinia, Sclerotium rolfsii,* and *Pellicularia*
- **Labeled crops:** many foliage, bedding, and flowering pot crops, including bulbs
- **REI:** twelve hours

PCNB is a contact fungicide with relatively long residual activity. Apply as a soil drench at seeding or transplanting. It shouldn't be applied through irrigation systems. Maintain agitation of spray solution to prevent settling and uneven application. Proper pathogen identification is essential if PCNB is used alone. Control spectrum does not include *Pythium* and *Phytophthora.* Most growers tank mix with a water-mold fungicide. Some bulbs may be dipped for control of *Rhizoctonia* and other rots. See label for specifics.

Thiophanate-methyl

- **Trade name:** Domain 4.5F & 50WP, Cleary's 3336 50WP 4.5F & 2G, Scotts Fungo WP and Fungo Flo
- **Pesticide class:** benzimidazole
- **Key diseases or pathogens controlled:** black spot of rose, *Botrytis, Fusarium, Phomopsis,* powdery mildews, *Septoria, Rhizoctonia, Sclerotinia,* and *Thielaviopsis*
- **Labeled crops:** most greenhouse and field-grown ornamentals
- **REI:** twelve hours

Broad-spectrum, systemic fungicide for foliar, drench, and dip applications. Flowable and WP formulations require thorough mixing and agitation. Apply with hydraulic sprayers. A recent study by Penn State

University revealed that many Pennsylvania greenhouses had thiophanate-methyl-resistant strains of *Botrytis.* Tank mixes with contact fungicides such as Protect T/O, Daconil 2787, or Chipco 26019 are helpful in reducing the chances of resistance and improving control if resistant strains are present. Numerous premixed combination products are available for foliar diseases including Benefit, Duosan, ConSyst, and Zyban. For all diseases, practice rotation with different classes of fungicides. For broad-spectrum root and crown rot prevention, tank mix with *Pythium/Phytophthora* selective fungicides. Banrot is a combination product containing thiophanate-methyl and etridiazole. If *Rhizoctonia* is active, however, it's better to apply thiophanate products alone as they're labeled for higher (curative) application rates than Banrot. Don't tank mix with copper-containing products or highly alkaline pesticides such as lime sulfur or Bordeaux mixture.

Vinclozolin
- **Trade name:** Ornalin 50WP & 4L, Curalan 50DF & 4L
- **Pesticide class:** dicarboximide
- **Key diseases or pathogens controlled:** *Botrytis* and *Sclerotinia*
- **Labeled crops:** many greenhouse and field-grown ornamentals (see label)
- **REI:** twelve hours

A contact fungicide with localized systemic activity. Excellent control of *Botrytis,* but exclusive use promotes resistant strains. Practice rotation and/or apply in tank mixes with fungicides in different chemical classes. Don't use exclusively in rotation with Chipco 26019 since the active ingredients are in the same chemical class. Apply as a spray or through thermal foggers. Maintain agitation of spray solution—even when using foggers. Not labeled for drench applications. Don't spray seedlings until they've formed three true leaves.

References

Hudson, William G., Melvin P. Garber, Ronald D. Oetting, Russell F. Mizell, Ann R. Chase, and Kane Bondari, 1996 Pest Management in the United States greenhouse and nursery industry: V. Insect and mite control. *HortTechnology* 6:200-206.

Jones, Ronald K., Ann R. Chase, Melvin P. Garber, William G. Hudson, Jeffrey G. Norcini, and Kane Bondari, 1996. Pest Management in the United States greenhouse and nursery industry: II. Disease Control. *HortTechnology* 6:200-206.

"1997 Cornell Recommendations for the Integrated Management of Greenhouse Florist Crops." Published by Cornell Cooperative Extension, Ithaca, NY.

"Tips on the Use and Safety of Chemical, Biologicals and the Environment of Floriculture Crops." Published by the Ohio Florists Association, November 1995.

November 1997.

Warning: Don't Violate the Label

Chris Beytes

We were recently reminded that while research articles published in *GrowerTalks* and other trade publications often evaluate various chemical treatments on plants, complete with recommended application timing and rates, unless a chemical is specifically labeled for a use, you risk violating the law if you make the same applications in your commercial greenhouse.

A case in point is the article on improving lily quality (*GrowerTalks*, March 1998, p. 82), that shows excellent results using spray applications of the hormones Promalin and Accel. However, neither of these products is registered for use on lilies or any other ornamental crop. Gibberellic acid (GA) has been shown also to minimize leaf yellowing of lilies but it is not labeled for lilies either.

Terril Nell, University of Florida floriculture professor and *GrowerTalks* contributing editor, mentions a case in Florida where an inspector found GA in a storage area a couple of years ago at a *Spathiphyllum* production range (GA wasn't labeled for use on that crop). The grower was told to get rid of the GA (i.e., don't use it any longer) or face fines. In the case of GA, the University of Florida and the Florida Nurserymen and Growers Association found a company willing to label GA for use on spaths and azaleas.

The bottom line is, research articles are intended to show what can be done to improve our crops. When chemicals tests show favorable results, the industry should push manufacturers to seek registration for that particular use, as Florida did with GA. Otherwise, as Terril suggests, seek chemical-free ways to improve your crops or at least work with registered products. For instance, in the case of lily postharvest quality, Terril points out that quality can be greatly improved through cultivar selection and avoiding storage.

GrowerTalks in Brief, May 1998.

Step-by-Step Safety

Brent Bates and Bob Decker

Safely storing and handling chemicals is paramount in the green industry, where losses for pollution-liability claims can climb into the hundreds of thousands of dollars, without even including legal expenses. That's why it's so important to carefully examine the way chemicals are stored and handled at your business and to take the necessary steps today to prevent losses.

Employee Safety, Training

Your employees are key in preventing losses due to improper chemical handling. Investing in training and equipment to help ensure their safety will ultimately benefit your business by reducing the likelihood of injuries or damages. Some employee-related safety measures include the following.

Educational safety classes

Consider mandatory training and safety awareness classes for all managers and employees, with regular refresher classes. These are often conducted through county or state extension offices.

Personal protective equipment (PPE)

Safety gear is vital for keeping employees safe from harm while working with potentially harmful chemicals—and it's required by law! Without PPE, the organophosphates in some pesticides can cause malfunction of red blood cells if they're inhaled or absorbed through the skin. If ingested, organophosphates can interfere with cholinesterase, an enzyme in the blood that works to ensure proper functioning of the nervous system. Over the last seven years, manufacturers have phased out organophosphates, but they're still present in some pesticides.

Federal law requires that all PPE specified on a chemical label be worn when applying the material. Be sure employees read and follow those instructions prior to working with chemicals.

Postapplication safety

Insist that employees take precautions after working with chemicals. Some postapplication safety tips for employees include:

- Always washing hands and face (including ears and nose) thoroughly immediately after handling chemicals and before smoking or eating
- Changing clothes after any contact with pesticides, fertilizers, or other toxic materials
- Thoroughly washing face shields with soap and water before they're worn by another person

Transport and Storage

Use only experienced drivers for transporting pesticides. Check their driving record for any serious violations or accidents. Also, inspect your vehicles before each trip to make sure they're in good working condition.

Store pesticides in a separate building. If this isn't possible, locate the storage area on an outside wall of the building, making sure it's not adjacent to an office or break room.

Make sure the storage area is well ventilated. Fans that turn on with a light switch are a good idea. Be certain that the floor is "liquid tight" and that joints between floors and walls are sealed to prevent leakage.

Use shelves constructed of metal or plastic, or cover shelves with an epoxy paint or plastic sheet. Build lips on the front of shelves to prevent bottles from falling off. Store heavier containers on lower shelves.

Don't store fertilizers near pesticides. Most fertilizers are oxidizers, meaning they can support a fire even if oxygen is cut off. They'll continue to burn even when water or other extinguishing materials are applied. If fire officials know that pesticides are involved in a blaze, they may not enter the building or even go near it to fight the fire.

Don't store PPE with pesticides or fertilizers.

Mixing Safely

Construct a pesticide mixing area that will contain any spills. If this isn't feasible, install a drain in the floor leading to a sump for collecting chemicals.

Install a fan in the mixing area that will pull any dust or vapors away from workers and exhaust them outside. The outside vent should be positioned to avoid blowing on people outside the building.

Be certain the mixing equipment is in good repair and that shaft seals are tight so they don't leak pesticides.

As a safety precaution, set up an emergency eyewash and shower station near the mixing area.

The High Cost of a Spill

In addition to the dangers posed to personnel within a facility, you must also be sure to take precautions to prevent toxins from being released outside your facility. Spills or fumigation leaks near a populated area can expose you to enormous costs in trying to defend medical claims, not to mention the exorbitant legal fees associated with multiple claims. In addition to legitimate claims that result from pesticide fumes or spills, these situations are typically ripe for fraud, as unaffected individuals jump on the claims bandwagon. It can be difficult to weed out fraudulent claims because all claimants' complaints are often subjective and/or vague, making defending your company difficult. Costs can go even higher when there's a government response to pollution claims. Settling claims with the Environmental Protection Agency (EPA) can be a lengthy process, and charges are usually five to ten times higher than with private remediation.

To protect your business from pollution losses, ask your insurance agent if pollution liability is covered under your current policy. GreenPack, marketed by Florists' Mutual and member companies of the American International Group Inc., is a pollution-liability insurance program designed to meet the needs of the horticulture industry. The basic GreenPack program includes:

- Coverage at owned or operated facilities: pays losses for pollution conditions emanating from within your location that result in bodily injury, property damage, or cleanup costs outside of your location. Coverage also is provided for on-site pollution conditions.

- Coverage at customers' sites: pays for losses for pollution conditions caused by application or overspray of herbicides, pesticides, or fertilizers on property that you neither own nor rent and that result in bodily injury or property damage.
- Coverage for owned or non-owned transportation: protects you against liability for bodily injury, property damage, or cleanup costs arising from spills that occur in the course of transit.
- GreenPack also offers underground storage tank coverage as an option. By law, owners and operators of underground storage tank systems must prove they're financially capable of managing a leak from a tank.

The standard GreenPack program offers limits of $1 million in coverage with a minimum deductible of $5,000 and a minimum premium of $2,500.

Put Safety First

It's vital that you protect your business against losses from improper handling and storage of chemicals. By training your workforce, exercising caution, and purchasing a pollution liability policy specific to your needs, you can do much to prevent costly losses down the road.

April 2000.

Lost and Found

Richard Lindquist

Regardless of which side of the pesticide fence you're sitting on, these are interesting—but hardly satisfying—times. If you're totally against the use of pesticides in pest management, well, they're still around and likely will remain so indefinitely. If you're a charter member of the "spray and pray" denomination and are especially fond of pesticides in the carbamate and organophosphate families, you'll probably see some of your favorites disappear from the market in the next few years.

The bottom line is there's enough going on in the pesticide area to upset nearly everyone, but enough positive changes to give everyone hope. Here's a summary of the pesticides our industry has gained and lost as well as a look at what might be ahead.

The Bad News

Government pesticide regulations aren't the only reason why a product might disappear. Pesticides can come and go from the greenhouse market—

or any market—for a number of reasons, including the product not being effective or causing phytotoxicity; a potential safety problem for applicators, workers, and customers (whether due to product misuse or not); marketing decisions by the manufacturer; and registration decisions by Environmental Protection Agency (EPA). Right now, however, pesticide manufacturers are dealing with at least two related things regarding pesticide registrations: reregistration and the Food Quality Protection Act of 1996 (FQPA). How are these related, you ask? Let me explain.

All pesticides that were registered before November 1, 1984, are required to be reregistered under provisions of the Federal Insecticide, Fungicide, and Rodenticide Act (FIFRA), which greatly increased testing and safety standards over previous regulations. Further, to be eligible for reregistration (or even registration for new products), pesticides must meet the safety requirements of the FQPA. This act requires additional safety margins for children and considers the total exposure to entire classes of pesticides when making decisions on registrations (the so-called risk cup). If the EPA deems that the pesticide class exposure safety margin is exceeded, the pesticide manufacturer must find ways to reduce exposure, either by canceling some (perhaps all) uses, reducing application rates, reducing the number of permitted applications, or increasing restricted-entry intervals (REI).

Because of the number of products in each category still registered and their widespread use in all types of pest control, the organophosphate and carbamate pesticides are the first to undergo the double whammy of reregistration under the FQPA requirements. More on that later.

During the past ten years, before the effects of FQPA have even started to affect the number of registered products but during the regular reregistration process, some key compounds and active ingredients have disappeared from the greenhouse market (but not necessarily from other uses) in the U.S. These include Methoxychlor, Lindane, Pentac, Lannate, Vydate/Oxamyl, Malathion, Vendex, Resmethrin, Sumithrin, Topcide, Margosan-O, and Ornamite.

I don't know the reasons for the removal of all of these pesticides, but I can take educated guesses. Lannate, Vydate/Oxamyl, and Vendex had their ornamentals registrations canceled by the DuPont Company as a direct result of the Benlate lawsuits. Lannate and Vydate/Oxamyl are carbamate products and would eventually undergo scrutiny under FQPA anyway. However, my understanding is that DuPont didn't consider the risk/reward ratio for these products on ornamental plants favorable.

Methoxychlor, Lindane, and Pentac probably disappeared because they're in the chlorinated hydrocarbon chemical family, a politically incorrect group of pesticides (relatives of DDT), and are products that would be difficult to reregister even without FQPA. Kelthane and Thiodan, which are still registered, may soon meet a similar fate. I believe that Ornamite's registration was also canceled because it would be difficult to reregister. Resmethrin and Sumithrin, both pyrethroids, probably disappeared because they were no longer effective. Topcide, also a pyrethroid, was OK as a pesticide, just not OK enough to outcompete existing products. Other azadirachtin products, such as Azatin, basically replaced Margosan-O.

As mentioned, the organophosphate and carbamate insecticides are the first groups to be reviewed under the risk cup approach. As a result, there have been some additional casualties. Products that have only a short time left on the market (the decisions have already been made) include sulfotepp (Dithio/Plantfume 103) and bendiocarb (Closure [=Dycarb]/Turcam). These have been voluntarily canceled by their manufacturers and will be phased out over the next few years.

Products "at risk" include all of the remaining organophosphate and carbamate pesticides. At the present time, chlorpyrifos (the active ingredient in DuraGuard) and acephate (the active ingredient in Orthene) are undergoing review. As this is being written, no decisions have been made.

Cancellation of a registration is one option. Another method of "risk mitigation" (to reduce the exposure, keep the risk cup from overflowing, and avoid canceling a registration) that EPA might require could be a five- or ten-day REI for a product. Although this wouldn't actually cancel a product, an extremely long REI will effectively prevent it from being used.

The Good News
Fortunately, there are some excellent new products in entirely new pesticide classes that have been registered for use on greenhouse crops. More are in the

registration process at this time and should appear on the market in the next few years. You already know and use many of the new products. They include Azatin and Ornazin (both contain azadirachtin), Avid and Conserve (both derived from soil microorganisms but not in the same chemical class), Endeavor (a systemic insecticide not related to Marathon), Marathon (nicotinoid systemic insecticide), Triact (neem oil, but without much azadirachtin), Botanigard/Naturalis-T&O (formulations of the fungus *Beauveria bassiana*), Cinnamite (cinnamaldehyde), Distance and Adept (both insect growth regulators), Sanmite, and Floramite.

When we look at the list of new products in our arsenal, it's clear that, so far, there's been a net gain in the number of pesticide products available and a wider range of pesticide classes to use. Most of them are very effective. This is indeed good news.

Does the development of all of these new products mean that it's now OK to get rid of the older organophosphate and carbamate pesticides? Not necessarily. The great thing about retaining a wide range of pesticide classes containing effective products is in implementing a pesticide rotation plan that doesn't overuse any single class.

So while it's wonderful to see the new pesticides being registered, if we're going to be forced—because of the registration cancellation of the older products—into repeatedly using only a few of them, the resistance management process won't improve.

April 2000.

Update from the Field

Lynn P. Griffith Jr.

Though the selection of agricultural chemicals available to growers has decreased steadily over the years, manufacturers continue to come out with new products that are better and safer than their predecessors were. Some of the products mentioned here might already have a place in your ongoing insect and disease management programs.

One exciting new fungicide is Heritage (azoxystrobin), from Zeneca. This fungicide probably has the widest range of target pathogens of any fungicide I've ever seen. It's effective against an extremely broad range of foliar and soil-borne pathogens. Some popular uses include control of *Phytophthora* and southern blight—pathogens with very limited control options. Heritage has also been effective against *Fusarium* as a drench. This is especially useful since some fungicides used to control *Fusarium* are either no longer effective or no longer available. The price per pound is high, but the rates of 1 to 4 oz. per 100 gal. are very modest. Growers have raved about its effectiveness, though everyone is urged to follow the label cautions and to use Heritage in rotational and tank mix programs to avoid resistance. In Latin America, the product is sold as Amistar.

Another fungicide, Decree, has been popular with growers for *Botrytis* control. Manufactured and marketed by SePRO Chemical Company, Decree is registered for *Botrytis* control on poinsettias, geraniums, and numerous bedding plants. A few foliage crops are also on the label. Decree has a four-hour REI (restricted-entry interval). Growers needing to control *Botrytis* on ornamentals should consider adding this fungicide to their program.

ConSyst from Regal Chemical isn't particularly new, but it seems to be growing in popularity in certain parts of the country. It's a mixture of two active ingredients, chlorothalonil (Daconil, Bravo) and thiophanate-methyl (Cleary 3336, Domain). While these active ingredients have been widely used for years, this combination gives ConSyst broad-spectrum control. I would caution growers about tank mixing other materials with ConSyst, however.

For insects and mites, Conserve SC from Dow AgroSciences has been growing in popularity. It controls a broad range of lepidopterous larvae (worms), along with western flower thrips, some leafminers, and two-spotted spider mites. Conserve is relatively easy to work into IPM programs, as it's gentle on beneficials, including predator mites. The four-hour REI is also convenient. In the past, Conserve SC was labeled only for use in lath and shade houses; it's now registered for greenhouses as well.

Growers are finding that the use of the pathogenic fungus *Beauveria bassiana* (Naturalis-T&O, BotaniGard, Mycotrol) in combination with insecticidal soaps is very effective. It seems the insecticidal soaps help breakdown the chitin (the waxy outer coating on the insect), thus allowing *B. bassiana* to enter more efficiently. This aids greatly in whitefly and mealybug control.

An older chemical that's becoming more popular is Hexygon, especially when tank mixed with other materials such as Avid. Its active ingredient, hexythiazox, is effective on spider mite eggs, and its use in combination with other miticides helps to extend the spray interval. This miticide from the Gowan Company not only controls mite eggs, it also affects adult female mites by rendering their eggs unviable. Since the miticides on the market aren't very effective against spider mite eggs, using Hexygon in combination with other miticides helps keep mite populations from coming back. Hexygon is effective against southern and European red mites, as well as two-spotted mites, and others. Gowan says they've applied for greenhouse registration for this product, so stay tuned.

Triact 70 is a multipurpose pesticide from Olympic Horticultural Products that's been reformulated from a 90% to a 70% product, with improved performance and greater odor control. Triact is a neem seed oil byproduct with little azadirachtin, so it's different from many of the other neem-based products. It has worked very well controlling powdery and downy mildew in bedding plant and greenhouse crops. It also works as a miticide, offering both contact and ovicidal activity. Phytotoxicity has been quite low, but as with any oil, be careful when temperatures exceed 85°F. Open blooms of some sensitive varieties such as impatiens may be affected, but otherwise Triact appears to be quite safe. It also has some insecticidal properties and sports a broad greenhouse and ornamental label, with a four-hour REI.

Growers who've been around a while have seen agricultural chemicals come to the marketplace and do a great job for a few years, but then their efficacy deteriorates as pests gain resistance. The chemical manufacturers are doing a better job writing chemical labels to inform growers about how to use rotations and tank mixing to help reduce resistance. It's more important today than ever before that you rotate chemistries and observe pest life cycles to help us keep these effective products around for a long time.

Pest Control, November 1999.

Greenhouse Chemicals: Trial, Then Trust

P. Allen Hammer

As you've no doubt heard by now, poinsettias have been removed from the Adept label. This is a real concern for me because we've effectively used Adept on poinsettias to control fungus gnats and shore flies for the last two years and have experienced no phytotoxicity from the chemical in our trials.

So why were poinsettias removed from the label? Several growers, particularly on the East Coast, reported a leaf burn and yellowing after applying Adept. Yet the reports I've heard say the phytotoxicity occurred because of overapplication of the chemical—in the range of five times or greater than the label rate!

Granted, the application rates and instructions for Adept are somewhat different from many of the pesticides we use in the greenhouse, but to me the label instructions are very clear. Adept shouldn't be applied as a heavy drench because the amount of chemical applied per container is very important. Many growers have been applying the chemical with a heavy drench application method, paying little attention to the amount of chemical applied per container. Nearly any greenhouse chemical would cause phytotoxicity or even plant death if applied at five times the label rate. Even fertilizing with 1,000 ppm of nitrogen could be extremely harmful to many crops. It's unfortunate that Adept has gotten such a bad reputation simply because it's new and needs to be used differently than most of the other pesticides we presently use.

Symptom of Our Times

Removing poinsettias from the Adept label also reflects our present business climate with respect to product liability. Our society increasingly assigns blame for our mistakes to others—in this case to the chemical. But if we blame phytotoxicity on the chemical when the chemical was improperly applied, we'll see fewer and fewer chemicals labeled for greenhouse crops. We simply can't afford to lose chemical labels through misuse of the product.

Some might suggest that we develop foolproof chemicals and instructions for use. However, given the large number of plants, variable greenhouse environments, and grower interpretations, I think it's almost impossible to completely eliminate the risk in chemical application. Even with the very best directions, we still tend to make our own interpretations of the label. However, applying Adept at five times the label rate is a dumb mistake, not a simple misinterpretation.

Start Slow

Our tendency to overzealously jump to new things without adequate trials is another industry trend that concerns me. Many growers will switch their medium, fertilizer, crop cultivar, pesticides, and other inputs in an entire greenhouse practically overnight, with little apparent concern for the large risk of such a rapid change. In no way am I suggesting we shouldn't change and trial new things. The key word is *trial.*

Personally, I think change should be a little slower. Admittedly, it's difficult for me to suggest that we're changing too rapidly when fifteen years ago I would have suggested our industry was much too slow to change. Now we've joined the rest of the world in embracing change. However, rapid change comes with a price tag. I think we need to be careful in deciding who must accept and carry the risk of change. If all the risk is assigned to a new product, we may reduce our opportunities for new technology.

So how should you use new chemicals? First, always carefully read and follow the label directions completely. Second, I'd suggest you trial any new chemical on a small scale before applying it to your entire crop. I never recommend a chemical treatment to a grower if I haven't used or studied the chemical. You simply can't afford to risk an entire crop.

Third, run your trial for an extended period before assuming everything's OK. With most pesticides, we've generally suggested that phytotoxicity will show up within a few days of application. However, this hasn't been the case with the Adept, which has added to the frustration. The phytotoxicity can appear long after the overapplication of the chemical.

And fourth, carefully record every chemical application with a lot more information than required by law. Be very specific: Record exactly how much chemical was mixed with a specific volume of water. Record when and how the chemical was applied, as well as how much spray or drench was applied to how many plants. Record the greenhouse environment and the weather before, during, and after the application. Without such information, it's often very difficult to reconstruct what went wrong. And remember, when a treatment is successful, you'd also like to be able to repeat your success.

Don't ever be afraid to trial new chemicals, cultivars, or techniques. A trial, however, not an acre or two of plants. A trial is the number of plants you're willing to risk in order to learn.

Growing Ideas, September 1999.

Success with Smokes/Fumigants and Aerosols

Raymond A. Cloyd

Though chemical pesticides are still the common method of controlling greenhouse insects/mites and diseases in commercial production systems, two other formulations— smokes/fumigants and total-release aerosols— are also widely used. These formulations are generally applied at the end of a crop cycle as a cleanup application. However, they're ineffective when pest populations build up and crop injury has already occurred. Smokes/fumigants and aerosols are generally used after hours when the greenhouse is closed and employees have left for the day.

Using Smokes/Fumigants

Smokes/fumigants come in cans that can treat 10,000 to 20,000 cu. ft. Cans are placed in the greenhouse walkway or elevated above the crop canopy. A sparkler or wire igniter is activated, triggering the release of the active ingredient.

Sparklers should be stored in a cool, dry place. Defective sparklers can fail to ignite the contents of the can. Be sure to check the sparklers before use, as they have a tendency to accumulate moisture. Don't use sparklers that are older than twelve months. When using smokes/fumigants, it's important to read the label for any precautions and guidelines for reentry.

Be sure to consult the label or call the manufacturer for ventilation precautions, especially when the greenhouse is situated in a residential area where public concerns might be an issue. Also, check to be sure that the smoke/fumigant that you're going to use is registered in your state.

When using smokes/fumigants, make sure the vents are closed and air circulation fans are shut off. It's also a good idea to notify your fire department that you're using a smoke/fumigant to prevent any false alarms.

Several insecticides/miticides and one fungicide are commercially available as smoke/fumigant formulations. These include Plantfume 103 and Dithio (sulfotepp), Nicotine Smoke (nicotine alkaloid), Thiodan Smoke (endosulfan), Vapona Smoke (dichlorvos), and Exotherm Termil (chlorothalonil). The insecticide/miticide formulations of smokes/fumigants are used against a wide range of pests including whiteflies, aphids, thrips, mealybugs, mites, and scales. These materials can penetrate flowers, petals, foliage, and dense canopies, which enhances their efficacy. Note that smokes/fumigants may be harmful to beneficial insects and mites. If you're using biological control in your greenhouse operation, be very careful when using these materials or avoid using them entirely.

Exotherm Termil is a fumigant fungicide that is used against *Botrytis cinerea*. The material is generally used before the onset of the disease. Cans should be evenly spaced at less than 50-ft. intervals and placed above the crop canopy. One can will treat 1,000 sq. ft. The REI is forty-eight hours.

Using Aerosols

The total-release aerosols are distributed by Whitmire Company Inc. Many active ingredients are available. These include fenoxycarb (Preclude), acephate (Orthene), bifenthrin (Attain), chlorpyrifos and cyfluthrin (Duraplex), and pyrethrin. Two-ounce cans that treat 3,000 sq. ft. and 6-oz. cans that treat 9,000 sq. ft. are commercially available.

Cans are generally placed above the crop canopy. The active ingredient in aerosols is released when a tab is pushed or activated.

When applying aerosols, close vents and run air circulation fans to help distribute the active ingredient among the crop and ensure thorough coverage. Aerosols may have limited movement when plants have very dense canopies. Aerosols are primarily used against flying insects such as adult fungus gnats, thrips, and whiteflies.

Pros and Cons

Growers prefer smokes/fumigants and aerosols because they're convenient and easy to use. Less protective equipment is required, and there is less exposure to the applicator than with pesticide application. However, applicators should leave the greenhouse immediately after activating these materials. Another precautionary procedure would be for two people to perform the application, with one individual suited up in protective clothing outside the greenhouse in case the individual performing the application is injured.

Pesticide effectiveness depends on having a good pest management strategy to maintain low pest numbers. So, when using smokes/fumigants or aerosols, it's imperative to have a pest management strategy that includes scouting, proper sanitation, and proper cultural practices. This can mean less pesticide use and better control of target pests, and it also helps to ensure a highly marketable crop for the grower.

Pest Control, November 2000.

Better Pesticide Management with Directed Sprays

Michael P. Parrella, Brook C. Murphy, Dave von Damm-Kattari, Gina von Damm-Kattari, and Chris Casey

One of the first steps in the development of an IPM program is for growers to learn to manage pesticides effectively. This is an especially important step for those growers interested in incorporating biological control into their programs. Unless growers have a handle on managing pesticides used in their operation, moving forward with biological control is probably a waste of time and money.

While many of the tenets of pest management are still in the research phase (that is, proper use of biological control, practical sampling programs, etc.), all growers can focus on better pesticide management just by using common sense. In fact, of all the pest management options a grower has available, techniques of pesticide management are probably the most widely adopted in our industry. This may be because it focuses on relatively minor changes to programs already in place by the grower—in essence, it's a slight modification of "business as usual."

Rotation Doesn't Reduce Use

One example of pesticide management is the concept of rotation of insecticides to delay the development of resistance in a particular pest. All growers are aware of this facet of IPM, as almost all growers have suffered from pesticide failure due to resistance development. Therefore, most growers place high priority on conserving the precious materials in their pesticide arsenal. And while the ways that products are rotated generally vary from grower to grower, this is probably the most widely used component of an effective IPM program and is an important pesticide management technique. It's safe to assume that growers are going to spray anyway; this just gets them to think a bit more critically about the sequence of what they're spraying and the consequences of relying on one or two materials.

However, while the concept of rotating pesticides changes what is sprayed, it doesn't reduce the amount of pesticide used or the number of applications made. This should be the goal of every grower. To accomplish this, other pesticide management techniques must be put into place.

Business-as-Usual Methods

For the past few years, with the majority of funding coming from the American Floral Endowment (AFE), we've concentrated on developing pesticide management techniques with the aim of reducing the amount of pesticide active ingredient used in greenhouses. These techniques can be divided into those requiring little or no modification of existing practices by growers and those requiring considerable modification (table 1). All the options listed in Table 1 should be considered by growers—the applicability of each will vary from grower to grower, but all are worth serious discussion. These options, if done properly, can reduce the amount of pesticide active ingredient applied and still provide the grower with acceptable control.

It's rare, however, for there to be any hard data on any of these options. An analogy would be the idea that keeping workers from wearing yellow clothing will reduce the number of whiteflies and aphids moving into a greenhouse. Since yellow attracts aphids and

Table 1. Management Options

Business as usual	Major modifications
Treat mother blocks before taking cuttings.	Use decision level sampling plan, leaf samples, sticky cards.
Dip/treat cuttings after taking them or upon arrival.	Spray only when needed based on thresholds.
Treat plants when still small (e.g., in trays, when plants are pot to pot, etc.)	Time sprays based on pest lifestage.
Spot treat	Degree day models
Directed sprays	

thrips, fewer of these pests will be on workers' clothing if they wear a color other than yellow. This makes sense, but who's going to conduct a study to show that this is true? We're trying to develop some hard data to show that some of these management options for pesticides actually reduce the amount of pesticide applied and yet will provide acceptable control for the grower. Recently we focused our attention on the technique of directed sprays.

Directed sprays, proven on roses to be as effective against thrips as high-volume sprays, are examples of improved pesticide management that can save growers time and money.
Photo by Chris Beytes.

The underlying assumption that directed sprays will work assumes that the insect or mite isn't distributed uniformly within a crop, but rather is concentrated on a certain plant section. When the distribution within the plant is known, sprays can be directed only to the plant part containing the largest numbers of the pest. Covering this section of the plant with spray will most likely take less water (and therefore less pesticide) than the alternative, which is to wet the entire plant. This concept will best fit where plants are vertically stratified (such as cut flowers) and growers can direct sprays to take advantage of this stratification. It may be more difficult to apply directed sprays to potted plants, where it may not be possible to direct sprays only to one section.

However, even with a crop such as pot mums, knowledge of insect distribution on the plant can lead to more intelligent and effective use of pesticides. We found out many years ago that melon and green peach aphids distributed themselves differently on potted mums. Melon aphids tended to concentrate near the tops of the plant, whereas green peach aphids would be more uniformly distributed over the plant. Armed with this knowledge and knowing the species of aphid on a crop, a grower could manage insecticides more widely. Obviously with melon aphids, making sure the central growth point is thoroughly covered with spray may be enough for effective control. In contrast, a full plant spray (including covering those difficult-to-reach lower leaves) may be necessary to control green peach aphid. Our recent work has focused on roses and an understanding of how thrips distribute themselves vertically in a rose crop.

Directed Sprays on Roses

We've been working with several cooperating rose growers in California, taking leaf and bud samples from the top to the bottom of each plant on a weekly basis, then examining the samples for the presence of western flower thrips. We've done this through several seasons of the year and with some of the most popular cultivars.

The first thing we noticed was that very few thrips were collected from the bottom parts of a rose plant. Most of the thrips were taken from the break, with most of those thrips concentrated near or on the developing bud (table 2). While we did find that certain cultivars were preferred by thrips over others, there was a tendency for thrips to occur near the developing bud regardless of cultivar.

With most of the thrips near the developing buds and few, if any, on the lower parts of the crop, directing sprays to the developing rose buds should provide control that is as good (or superior) to drenching the entire plant. Intuitively, this makes sense, but we set out to prove this through experimentation.

Table 2. Thrips Populations Found at Various Levels of Cut Rose Plants				
Crop level	**Larvae**	**Males**	**Females**	**Total thrips**
Top	0.33	0.11	0.53	0.87
Middle	0.21	0.03	0.16	0.46
Bottom	0.20	0.05	0.15	0.41

We tested the concept with a cooperating grower who used conventional pesticides at label recommended rates in a replicated greenhouse experiment. We asked the grower to make a normal application of each material to the crop and then make an application only to the developing buds. The directed sprays and full volume sprays used approximately 150 and 500 gal. of water per acre, respectively.

After the experiment, an analysis of the number of thrips in developing buds revealed little significant difference in thrips control between the high volume and directed sprays (fig. 1). And we achieved this equivalent control using nearly 65% less water and pesticide.

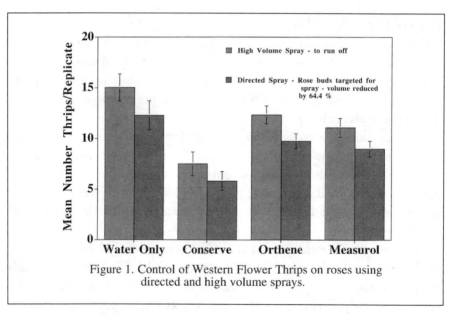

Figure 1. Control of Western Flower Thrips on roses using directed and high volume sprays.

Figure 1. Comparing directed and high-volume sprays.

Good Practice for the Future

With this data, we believe that rose growers who use a directed spray can achieve thrips control equivalent to their standard full volume wet sprays. They'll save on active ingredient, have few problems with runoff, and contribute to their goal of managing pesticide resistance in the greenhouse—all in one simple action. In addition, by directing sprays toward the top of the plant, they may be opening up the prospects for biological control of mites in the lower canopy.

With stricter worker reentry intervals, development of pesticide resistance, advent of new pests, and the full impact of the Food Quality Protection Act still to be felt, adopting IPM programs may be the only way for growers to survive. These programs are slowly coming online in the U.S. We believe that techniques like directed sprays, which require only a slight modification of grower practices, may speed adoption of IPM in floriculture.

January 1999.

Seven Ways to Maximize Sprays

Kurt Becker

One of the least-liked and most time-consuming jobs in the greenhouse is spraying to control disease and insects. Unfortunately, it's also one of the most important. But with some common sense and a few simple techniques, this arduous task can become easier and more effective. Here are seven tips for more effective chemical applications.

Tip One: Choose Your Tools Carefully

There are many different methods to spray your crop, from using hand-pump sprayers to automatic low-volume applicators. Each has its strengths and weaknesses and its place in the greenhouse.

Typically, the sprayer world is divided into two camps: high volume and low volume. These terms refer to the amount of water applied as a carrier. High-volume sprays are generally wet, employ particles larger than 100 microns in size, and are targeted directly at the plants. (A micron is equal to 0.0001 cm or 0.000039 in. One hundred microns is roughly the width of a human hair.) Low-volume sprays generally keep the plants much drier because of the smaller particle sizes. They can be either targeted directly at plants or used to treat a space, such as a bay or a greenhouse.

High volume and low volume each has its use in effective pest management. High-volume hydraulic sprayers are effective spot-treatment tools and are useful when the chemical being applied needs more water to be effective. They're also well suited to controlling certain pests, such as high infestations of scale or mealybugs. Targeted low-volume sprayers are most often faster than standard hydraulic sprayers, as they use less

Smart growers have low- and high-volume sprayers on hand to combat any pest problem.
Photo by Chris Beytes.

water as a carrier. They create smaller particles and often cover more surface area at greater distances. Low-volume sprayers are flexible, offering the ability to treat areas selectively. They can even be used in shade houses and outdoors, provided there's no wind.

Thermal foggers are the fastest method of chemical application. They can treat large areas and cover great distances as much as fifty times faster than high-volume sprayers. Thermal foggers create ultrafine particles (10 to 20 microns) that can remain airborne for several hours, enveloping the plants in a fog and covering both upper and lower leaf surfaces.

Automatic aerosol generators work similarly, creating billions of microparticles (5 to 10 microns), but they do this more slowly. The advantage is that they operate automatically; no operator is needed during the application. But because no one is there to direct the spray, proper machine placement and air movement are critical to ensure maximum coverage. Both the thermal foggers and the automatic aerosol generators must be used in enclosed spaces and can't be used selectively for spot treatments.

Which should you use? Your choice should be based on the crop, the crop stage, the pest, the size of area to be treated, the timing of spray, the time needed to apply, and the type of chemical you're using.

No one sprayer will do every job. For best results, you should own several types of spray equipment. View your sprayers as tools in a toolbox, with each having its individual use. A thermal fogger is a good choice for spraying one acre of bedding plants, while a two-gallon hand-pump sprayer would be better suited when your only pests are five aphids in the corner of the greenhouse.

Tip Two: Size Matters

Droplet particle size is an important consideration when applying pesticides. Standard hydraulic sprays vary in coarseness from 100 microns to more than 500 microns. While the term *low volume* refers to the amount of solution applied to a given area, particles from targeted low-volume sprays are typically less than 100 microns, and particles from aerosol low-volume sprays are generally less than 25 microns.

Generally, the finer the spray and the smaller the particle size, the more surface area that spray will cover. Because the mathematical relationship between particle size and volume is a cubic one, reducing the particle size by 50% increases the number of droplets available to cover plant surfaces by approximately 800%. For example, when you reduce the particle size of a spray droplet from 100 microns to 50 microns, each 100-micron droplet yields eight 50-micron droplets. These eight droplets are able to cover a much greater surface area than one 100-micron droplet. This relationship explains how low-volume sprays can cover large areas with less water.

Here's a good example: One 100-micron particle can be separated into 1,005 10-micron particles. If you were an aphid, a droplet of pesticide solution measuring 100 microns in size would seem like a small boulder to you, while 1,005 droplets would be like gravel under your feet. You could walk around the boulder, but you couldn't avoid the gravel.

Additionally, particle size is an important consideration in controlling runoff. The smaller the particle, the less effect gravity has on it. This, combined with reduced volumes of water, means less runoff.

Particle size has a great effect on hydraulic sprays as well. You can get eight 100-micron droplets from one 200-micron droplet. A 100-micron droplet is smaller and has less mass, which allows it to be more affected by the turbulence of the spray and less by the spray's forward inertia. The more swirling in the spray, the greater the under-leaf coverage. While the science behind these techniques may be complicated, the results aren't: Smaller particle sizes mean more area covered with less water and less runoff.

Tip Three: Train Yourself

Learning how to properly use each sprayer in your arsenal will greatly affect your results. For instance, there's a big difference in spray technique between using a standard high-volume machine and a low-volume machine.

Several commercially available tools can help you gauge your spray coverage with each type of equipment. Fluorescent dyes sold by Day-Glo Color Corporation, Cleveland, Ohio, (216) 391-7070, or hydrosensitive paper available from The Dramm Corporation, Manitowoc, Wisconsin, (800) 258-0848, are excellent tools for determining your spray coverage. By using a dye in your tank or by paper-clipping small pieces of hydrosensitive paper throughout the crop, you're able to see where you're doing a good job spraying and where your technique needs help.

In general, the dye is best for ultra-low-volume sprays, as the spray droplets are often too small to register on the hydrosensitive paper. However, dyes can be messy. For best viewing, use Blaze Orange dye and a black light to check your coverage.

When training with hydraulic or targeted low-volume equipment, the hydrosensitive paper is a less colorful way of determining coverage. It gives excellent results and requires less cleaning. Simply use a paperclip to attach bits of the paper to various plants, both above and under the foliage. After spraying, the paper will show if you covered the plants thoroughly or are missing areas when you spray.

Tip Four: Time Yourself

You need to know the output rate of any sprayer you're using. By combining this with the volume of solution you're going to use in a given area, you can even out your pesticide coverage throughout the house.

For example, if you know that you're going to apply forty gallons of a high-volume spray to a greenhouse and you know that your equipment's output rate is one gallon per minute (GPM), you know that it will take forty minutes to empty the tank. Therefore, at ten minutes, you should have completed a quarter of the area to be treated; at twenty minutes, half of the area should be finished. Too often, either too much or too little solution will be left near the end of the spray job. This means that the end of the house will either get blasted or won't get enough coverage. By metering your application speed, you'll ensure more uniform coverage throughout the entire greenhouse.

Tip Five: Time Your Application

The time of day when you make your chemical applications can impact their effectiveness. You usually can't make ultra-low-volume sprays during the heat of the day because of the need to close the greenhouse and turn off all ventilation. Hydraulic sprays in the midday sun may burn the crop because of large droplet evaporation. Keep in mind the various factors that limit your spray applications and consider them when choosing equipment, chemical, and time of application.

Tip Six: Check Nozzle Accuracy

Many of the above tips, such as understanding droplet size, depend on accurate nozzle output. Over time, nozzles wear due to abrasion and erosion, increasing the size of the orifice and changing output, spray pressure, and particle size.

For instance, imagine that you're applying twenty gallons to an area and are carefully metering your application speed. Your sprayer usually applies 1 GPM, but because of nozzle wear it's now spraying 1.5 GPM. It won't take long before you're rushing to finish when you realize that you don't have enough solution left to complete the rest of the house. Plus, your droplet size has also gotten larger as a result of this nozzle wear, reducing your spray coverage.

Because of this, it's very important that you regularly check your nozzles for accuracy and replace them when worn.

Tip Seven: Know the Enemy

In *The Art of War*, Sun Tzu wrote, "If you know the enemy and know yourself, you need not fear the result of a hundred battles. If you know yourself but not the enemy, for every victory gained you will also suffer a defeat." Sun Tzu must have known something about whiteflies.

It's important to properly identify the pests you're battling—through scouting, sticky cards, and greenhouse-personnel training—so that you are able to time your applications to take advantage of their weaknesses. For instance, you can increase or decrease the frequency of applications depending on their life cycle, target your applications to the pests' home through sprayer choice and technique, and choose the controls that will have the greatest effect.

Spraying the upper-leaf surfaces when hunting for aphids will get you more kills than spraying just the lower foliage. Likewise, making a spray application when it's still warm in the greenhouse may help you bag a few more thrips, as they're more active and mobile in warmer temperatures. And spraying for flying insects is generally more effective using ultra-low-volume sprayers, as the chemical remains airborne for longer periods.

April 1999.

Hydrogen Peroxide: Crop Cure-All?

Chris Beytes

Grower Charles Sowders, C&G Greenhouses, Troup, Texas, made an offhand comment in August's "Under an Acre" about using hydrogen peroxide to kill snails, fungus gnat larvae, and to even stimulate plant growth. That comment didn't go unnoticed by sharp-eyed readers who wanted to know more, so we asked Charles for details.

He says he first learned of H_2O_2's agricultural applications through a research paper written by a University of Michigan student that made a small mention about uses on plants.

"We had twenty-three days of cold, wet February weather, and *Botrytis* was consuming my New Guinea impatiens cuttings [that were being rooted]. I decided to try the hydrogen peroxide on the *Botrytis* because the plants were going to die anyway." He says that the research paper listed a supplier, so he ordered fifteen gallons as a starter.

He sprayed the cuttings using a 1,000:1 dilution, and "the *Botrytis* disappeared and never came back," Charles recalls. "Within an hour the plants had perked up and looked healthier than even at the beginning. The next day new roots had grown about a quarter inch. The success rate of the cuttings was 100%."

Next he tested hydrogen peroxide on rooted geranium cuttings. Twenty flats accidentally received an application at a 100:1 ratio. "In three weeks, the twenty flats were three inches taller than the others. The root development was much better, probably the reason for the more vigorous growth." He says that as the chemical breaks down into water and oxygen, the extra oxygen stimulates root growth. Charles now injects it into all the water in his greenhouses.

We asked some industry experts about hydrogen peroxide and learned that Canadian growers have some experience with it, probably having picked up tips from Dutch growers. But is this legal? According to Charles, it is because H_2O_2 is composed of two naturally occurring elements, hydrogen and oxygen. He adds that courts have upheld its unregistered use. However, we've been warned that you can't deliberately apply anything to a crop as a pesticide if it doesn't have an Environmental Protection Agency registration number.

For safety's sake, try ZeroTol from BioSafe Systems, Glastonbury, Connecticut. It's a 27% solution of hydrogen dioxide that's labeled for use as a broad-spectrum algaecide and fungicide. Crops labeled include bedding plants, cactus, flowering plants, nursery stock, ornamentals, propagation material, roses, seeds, trees, and turf. Apply as a spray, drench, or dip. You can also use it for general greenhouse sanitation and in cooling pads to control algae growth. Best of all, it has a zero-hour REI. You can get more information at BioSafe's Web site, http://biosafesystems.com.

Culture Notes, November 1999.

Chapter 4
Culture and Operations

Simple Ways to Reduce Insect and Disease Risks

Raymond A. Cloyd and Clifford S. Sadof

Good horticultural practices are the foundation of any successful pest management program. Over- or underfertilization, using the improper medium, overwatering, lack of sanitation, or crowding can make it difficult for even pesticides to rescue a crop. Following are some easy cultural management strategies that will go a long way toward reducing the risk of disease, insect, and mite attacks on your crops.

Fertility

High or low fertility levels can increase crop susceptibility to diseases, insects, and mites. Many plants grown under optimal light and nutrient conditions produce chemical defenses that can protect them from diseases and insects. Changes in light and nutrition can compromise these defenses, giving pests a clear path to the crop. For example, chrysanthemums are more susceptible to *Erwinia chrysanthemi* and *Pseudomonas cichorii* under high fertilizer rates.

High fertility levels can increase soluble salts in the media, which stresses plants and increases their susceptibility to root rot diseases (*Pythium* and *Phytophthora*). The soft, succulent tissue associated with excessive fertilization is often easier for insects/mites to penetrate with their mouthparts.

In addition, plants respond to high fertility levels by moving more nutrients to new growth. This provides pests easier access to nutrients they need to grow and reproduce. Consequently, insects and mites can grow faster and cause greater injury to the crop.

Media

Growers who use well-drained media minimize problems with root rots, fungus gnats, and shore flies. Media that drain well have adequate air pore spaces and allow water to pass through easily. Poor-draining media hold too much water and contain less air space. This creates conditions that promote root rot development. Media that don't drain well can hold more water on the surface, promoting algae growth. Both fungus gnats and shore flies breed in algae. If this problem isn't corrected, high populations of both insects can build up in a short period of time.

Fungus gnat larvae damage plants directly by feeding on root tips and tunneling into roots and stems. This reduces plants' ability to take up water and nutrients. In addition, they have the potential to transmit *Pythium* and *Thielaviopsis*. Shore flies don't directly damage plants because they primarily feed on algae. However, large numbers can present a nuisance problem, and they have been implicated in disease transmission.

Watering

Overwatering plants predisposes them to root rot problems and increased numbers of fungus gnats and shore flies. Excessive watering stresses plants because excess water fills air-pore spaces, creating low oxygen levels. This increases susceptibility to root rots. High moisture levels keep the medium surface moist, resulting in algae buildup that provides breeding sites for fungus gnats and shore flies.

Sanitation

Weeds and plant debris provide sites that pathogens, insects, and mites can use to survive and spread to the main crop. Weeds harbor pathogens, most notably viruses that are obtained by insects and then transmitted to crops when they feed. Weeds that serve as reservoirs for tomato spotted wilt virus are chickweed, lambsquarters, nightshades, shepherd's purse, pigweed, and bindweed.

Also, some weeds, such as sowthistle (*Sonchus* spp.), are susceptible to powdery mildew, which can migrate from the weed onto the main crop. In addition to harboring diseases, weeds provide refuge for whiteflies, aphids, leafminers, and spider mites, which can then move from desiccating weeds onto the main crop.

Plant debris (i.e., leaves, flowers, and media) provides refuge for the resting stages of various pathogens, insects, and mites. Pathogens can be spread onto crops from dried plant debris that's subject to splashing water or air movement. *Erwinia chrysanthemi,* which causes chrysanthemum bacterial blight, survives in dead plant debris that serves as a source of infection to newly introduced plants. Insects and mites can migrate to fresh plant material as plant debris dries out. Also, leftover media provide sites for fungus gnat adults to lay eggs and western flower thrips to pupate.

Crop Spacing

Plants spaced closely together reduce light penetration to lower leaves, increase humidity, and allow leaves to stay wet longer. This increases susceptibility to diseases such as *Botrytis* and powdery mildew. Plants that are spaced too close together increase the possibility of insects and mites moving from plant to plant when leaves touch. In addition, plants spaced too close together make it difficult to get uniform coverage with foliar-applied insecticides or fungicides.

Greenhouse managers who adopt proper cultural practices for their crops are likely to spend less time fighting pest problems and more time growing and marketing a healthy crop.

Editor's note: The authors would like to thank Karen K. Rane, Botany/Plant Pathology Department, Purdue University, for her comments.

November 1997.

Managing Magnesium in Foliage

Dr. James F. Knauss

Magnesium, along with calcium, is classed as a secondary nutrient in the fertilizer world, but by no means should this be taken as a lower class status than the N-P-K part of the nutritional picture. In some areas, magnesium nutrition is more difficult to manage.

Although deficiencies of calcium can occur, these are less common than its secondary nutrient partner, magnesium. Calcium is usually abundant enough in water and in the soil mix, while magnesium is definitely less abundant. Deficiencies are seen more often in high-magnesium-demanding crops that are in container production for three months or longer.

Limestone Isn't Always Enough

Many growers believe that if they use dolomitic limestone, a combination of calcium and magnesium carbonate, in a growing mix, their magnesium worries are over. Not true! Although dolomitic limestone provides this necessary element, often forces beyond a grower's control produce conditions where magnesium supply may become limited.

If the limitation lasts for an extended time, plants respond by transferring magnesium from lower leaves to new growth, resulting in yellow, deficient lower leaves and normal levels of magnesium in new growth. Increasing dolomitic lime levels in the mix seems like a logical approach to solving the problem, but there is a limit to how much lime you can add before plants start showing adverse effects.

Growers often use Epsom salts (magnesium sulfate) to provide additional magnesium. This has been applied as a periodic foliar spray (usually about 5 lbs. per 100 gal. water), as a component in a soil mix (about 5 lbs. per 3 yds.) or as a periodic drench to the soil mix at 1 to 2 lbs. per 100 gal. water applied at 2 to 4 pints per 2 ft. of soil surface. We've found that when Epsom salts are applied to soil mixes, the magnesium component is very soluble, is often easily leached, and the available magnesium supply to the plant is usually short lived.

Other forms of magnesium, such as coated release forms, are available. Their costs to the grower and the magnesium availability to the plant vary considerably.

Availability of magnesium is affected by a soil mix's pH. Normally, pH levels between 5.5 and 6.5 produce an acceptable availability of magnesium. Levels of pH 7.0 and above can severely affect magnesium availability. When the pH level drops below the normal range, magnesium becomes very available and excessive leaching and loss can occur, producing deficient magnesium conditions.

What's the Appropriate Magnesium System?

Always use a high-grade dolomitic lime in your soil mix as one component in your magnesium nutritional management system. Insist on a dolomitic lime with a calcium carbonate equivalent (CCE) of 103 or above. This is a measure of the lime's purity, with 109 being the highest theoretical value, though it's never achieved in commercial lime sources.

Test your irrigation water through a reputable laboratory. You need to know how much magnesium relative to calcium and sodium (direct antagonists to magnesium uptake) is present. These ions accumulate in the soil mix, increasing with increasing frequency of irrigation. Where rainfall makes up a portion of irrigation, the rainwater must be viewed to be similar to distilled water (very little magnesium in it, with no moderating effect on soil pH).

One procedure to bring up the magnesium level in your water is injecting Epsom salts to provide supplemental magnesium every time you irrigate. If this procedure is used in nursery production, the amount of Epsom salts you add probably shouldn't exceed the level necessary to bring the calcium-magnesium ratio to greater than one 1:4. Adding more is usually unnecessary and increases your costs.

Magnesium Nutrition Musts

1. Always determine if magnesium deficiency is present by analyzing soil mix and appropriate leaf tissue from plants showing typical lower, older leaf chlorosis. Be sure to use a reputable laboratory, and consult with them on what samples to take.

2. Analyze your water through the same laboratory to determine what role it can play in the overall nutritional picture.

3. If magnesium is deficient, examine the quality and effective rate of the dolomitic lime used in the soil mix. You may need to change the source and increase the amount per cubic yard.

4. Select fertilizers with additional magnesium.

5. Where necessary, initiate foliar nutritional sprays on a fourteen- to twenty-one-day schedule using 27-15-12 or 20-20-20 added to 100 gal. of water at 3 lbs. Add 5 lbs. of Epsom salts to this solution.

Culture Notes, February 1998.

Quick-Crop Pest Control

Will Healy

During the hectic spring rush, controlling pests becomes a nightmare for growers because of short crop times, insufficient labor, and the need to be working in the crop continuously. If a pest problem appears, the control strategy must be efficient and effective. When developing your Quick Cropping Program, consider the following.

Clean Plants In, Clean Plants Out

Failure to inspect the plants in your holding area is the main reason growers end up spraying in the spring. Inspect all plants prior to transplanting to eliminate insects and disease before planting!

Applying protective sprays to the plants in your holding area will ensure that insect and disease problems are eradicated prior to transplanting. Once you transplant a problem crop, the cost of control grows exponentially and your success rate drops.

When moving transplants into a problem area, apply a protective spray to reduce the threat of insects and diseases. Use the newer chemicals that have translaminar or systemic activity to make sure the plants are thoroughly protected during the early days in production. For example, when transplanting impatiens into an area with known bacterial problems, apply Phyton 27 prior to transplanting to protect the plants during the first four to seven days after transplanting.

Aphids from Heaven

How many times have you sworn that the insect problem appeared out of nowhere? Just like bears, insect come out of hibernation in the spring. Many insects have an overwintering stage that allows them to survive through harsh conditions.

If your garden mums had thrips last fall, you'll find a thrips problem in the same area when the weather warms up this spring. Move thrips-tolerant plants (petunias) into these problem areas to minimize the need to spray. Avoid placing problem crops near high-risk areas to reduce your need to spray. If you must spray, consider using Conserve, which has a short REI.

Aphids can suddenly appear in large numbers outdoors and invade your greenhouse. Monitor plants inside and outside the greenhouse to prevent invasive pest problems. A regular spray program using horticultural oils on the new vegetation outside the greenhouse will control outside pests.

Singing in the Rain

If the weather changes to wet and dreary, be on the look out for increased disease problems. Within a matter of hours, a disease-free crop can become seriously damaged when the weather conditions turn nasty.

Monitor for the sudden appearance of black, blue, or red leaf spots due to *Alternaria, Pseudomonas,* or *Anthracnose.* Treat the plants as soon as spots appear—delaying treatment will allow a few spots to develop into extensive foliar damage. Heritage is a new product that's been effective at controlling the fungal leaf spots, while any of the copper fungicides will control the *Pseudomonas.*

Not only are foliar diseases a problem in rainy weather, but wet soil is a perfect environment for establishment of *Pythium, Thielaviopsis,* and *Rhizoctonia.* Check *Pythium*-sensitive crops (snapdragons, dianthus) for brown roots and treat with Subdue Maxx or Aliette if an extended wet period is in the forecast. Keep the soil pH below 6.2 to minimize *Thielaviopsis* problems.

Botrytis and mildew diseases are best treated environmentally. If you keep the humidity below 100% (the dew point), these disease problems won't occur. Many growers struggle to keep humidity below the dew point because they forget that warm air holds more water than cool air. Anytime the greenhouse temperature drops, the humidity increases. To stay under the dew point, increase the temperature and ventilate or reduce the application of water to the greenhouse.

There are also a number of new chemicals in the strobilurin class (Heritage, Cygnus, Compass) with low REIs that are effective at managing *Botrytis* and mildews. Check the labels for specific crop warnings. Avoid excessive spray treatment, since spraying adds water to the greenhouse environment, raising the relative humidity and enhancing the spread of the disease.

Pest Management, April 2000.

Cleaning up Recycled Water

Don C. Wilkerson

A well-designed, efficient irrigation system is the foundation of a good water management program. Obviously, the less runoff you have to deal with in the first place, the less of a problem it creates down the road. Drip and subirrigation do an excellent job of delivering water efficiently, whereas overhead systems create large volumes of runoff and increase the potential for disease and insect problems.

However, all recycling systems require some form of filtration or treatment. Here is a short summary of what's available today.

Filtration and Sedimentation

Filtration can play a very important role in reducing the risk from pathogens in a recycling system. Turbidity, algae, and sediment not only clog irrigation equipment and emitters, they also provide a source of harmful pathogens.

There have been numerous advancements in this area over the past five to ten years. These include self-backwashing filters, improved sand filters, sediment membranes, spindown filters, and low-cost, replaceable cartridges.

Netafim's spin clean disk filter system is among the most popular. This system provides filtration down to 130 microns and offers a self-backwash option based on pressure differential.

Although filtration can be expensive, you can use more passive methods of reducing sediment, such as grassways or weirs. Chemical flocculents have also played an important role in this process.

Disinfection

Recycled irrigation water may become contaminated with harmful pathogens such as *Phytophthora* and *Pythium*. However, disinfecting recycled irrigation water is difficult at best because of the complexity of most systems.

Chlorine gas has frequently been used as an oxidizing agent to kill organisms in solution. Typically 3 to 5 ppm, with a contact time of approximately three to five minutes, is sufficient to kill most pathogens. However, due to the many safety and health restrictions placed on chlorine gas, most growers no longer rely on this technique.

The latest approach uses an electrolysis process to convert potassium chloride to potassium hypochlorite. The chlorine from potassium hypochlorite works similarly to traditional chlorination systems but without the health and safety risks.

Ultraviolet (UV) light has also been used to treat irrigation water for pathogens. However, obtaining adequate levels and exposure time to UV light has been a limiting factor. The biggest problem occurs when large volumes of water pass through the system very rapidly. The UV light may not get adequate contact time to disinfect. This is compounded when turbidity is high because the UV light can't penetrate the solution.

Plant pathologists are now studying the effects of lasers on pathogens in solution. The biggest challenge to date has been recharging the system between laser shots to effectively treat large volumes of rapidly moving water. Again, the situation is compounded when turbidity is high.

Ozone is another oxidizing process that has been used on a limited basis. Although ozone has demonstrated an ability to destroy a broad range of harmful pathogens, there are still many questions about the potential effect on plant nutrients, as well as on the plant itself. Ozone generators are relatively inexpensive, but measuring ozone levels in solution can be pricey.

The use of eloptic energy (electrostatic precipitation, ESP) for treating irrigation water has focused primarily on mineral content. The claims seem almost too good to be true; however, there's little research to back them up. ESP uses a form of radionic energy that obeys some laws of electricity and some of optics. The system supposedly works by rearranging electrons in a water molecule, changing its ionic nature. Manufacturers claim that these systems help control a wide range of plant pests.

Residual pesticides and other contaminants can create potential problems in recycled water, but they're very difficult and expensive to monitor because many of these chemicals are measured in parts per trillion. Activated charcoal (AC) has been used to chemically filter recycled irrigation water. Because of its very high ion exchange complex, AC has the ability to bond potential contaminants to its surface. The downside to using AC is that it can quickly become saturated—all of the ion exchange sites become occupied, allowing potential contaminants to pass through the system. And saturated AC presents its own environmental hazard and must be disposed of properly or refired.

Researchers continue to look for cost-effective systems to address both the environmental issue of runoff and the challenges of using recycled water for crop production. Hopefully, new technology will resolve both issues in the near future.

Pest Management, January 2000.

Maintain Precise Records to Make Diagnosis Easier
P. Allen Hammer

Keeping detailed records of greenhouse crop production should be a number-one priority for all greenhouse growers. This isn't a unique or new concept, but it may be a revelation to many growers. The record keeping I am talking about goes way beyond a simple card or label in a pot on the bench, although some growers may consider that to be enough. We need a great deal more information. Record everything you do to the crop, with a date and time: The exact amount of chemical you mixed with the exact number of gallons of water, how many gallons of spray you applied and to how many plants, the greenhouse temperature and the outside weather conditions during the chemical application, times when treatment or fertilizer were applied, and every change in water treatment or fertilizer analysis.

Why Bother?

The reasons you need to keep crop production records are very simple. Records are often essential—at the least, they're very useful in diagnosing plant problems or pinpointing troublesome areas. We're often faced with trying to diagnose a plant's growth problem without any knowledge of what may have been applied to the plant. "I think we applied chemical x at the label rate a couple of weeks ago" isn't very helpful or informative. But a simple mathematical or mixing error can be reconstructed if you have the details of the mix you utilized. Also, if you grow an outstanding crop, specific records of what you did during production can help you repeat that success.

Although we think we can remember what, how, and when certain things were done in the greenhouse, our memories tend to quickly fade. If in the last month you've tried to remember when something was done in the greenhouse, I would suggest you need to improve your record keeping.

Start Records with Plugs and Cuttings

This brings me to another place in the plant record keeping argument. As our industry moves toward specialty cutting and plug producers, plant history is even more important. More and more I'm faced with plant problems from growers who can provide no history of the cutting or plug they purchased, often from an unknown greenhouse. I'm convinced that it would serve our industry well if the plug or cutting history were provided to the grower at purchase.

Many will quickly suggest that this is impossible and unnecessary, but I'll argue that this can provide valuable help to both the producer and the buyer of the cutting or plug. I'm concerned that the slow early growth on some cuttings and plugs I've observed is caused by the overapplication of chemical growth regulators or by plant stress from holding the cuttings or plugs before the grower receives the plants. If the producers of cuttings and plugs were required to provide the plant history, more attention and concern would probably be exercised in what treatments or care the plants receive before shipping. I certainly don't want to imply that every greenhouse problem is or should be blamed on the cutting or plug producer, but we could rule out suspected problems if we knew the total plant history. It would be very helpful to know when the cutting was taken or when the seed was sown. It's often helpful to apply growth regulators to cuttings and plugs, and it would be extremely valuable to the grower purchasing the plants to have that information.

Database Management

Growers often ask me how they should keep such records. I suggest using a computer database program. The database will give you the ability to easily search for specific information, as the recorded information can be sorted and printed in various forms.

I'm still waiting for the first computer plant history form to be included with a problem plant. The best I have received is a copy of a handwritten plant history. The major hurdle to recording plant information is simply starting to track it. Once database forms are developed, you'll find it very easy to record information. Small hand-held computers are also available. They could easily be used for such data recording and would even make it fun to record the information.

It's time we embrace new technology. Record keeping is not a fad, and we have great tools to help. Plant records will make you a much better grower, and they're essential in today's business environment.

Growing Ideas, April 1999.

Managing Weeds in Perennials

Joseph C. Neal

When it comes to weeds, "Start clean, stay clean" should be every grower's motto. This is especially true for perennial producers. Although we can control most grassy weeds with postemergence herbicides, we have few preemergence herbicides that are safe on perennials and no postemergence herbicides to use when broad-leaf weeds get out of hand. Furthermore, the preemergence herbicides labeled for use in perennials are either safe on many perennials but don't control many weeds or control lots of weeds but are safe on only a few perennials.

Managing weeds effectively takes a comprehensive program that includes exclusion, sanitation, preemergence herbicides, and hand weeding.

Exclusion and Sanitation

Weed seed and other propagules are introduced into nurseries in the potting substrates, by wind-blown seed, splashed into pots by rain, deposited by birds, and (perhaps most important) preexisting in the plant materials themselves.

You should inspect new shipments of liners before potting. If you observe weeds that aren't currently present at your nursery, you have two choices: Refuse the shipment or remove the top half-inch of potting media from the liners and dispose of contaminated media. Closely monitor new plants in the nursery to prevent introduced weeds from going to seed. And cull plants from the nursery that are infested with perennial weeds such as nutsedge.

The worst weed infestations are those that build over time. Weed frequently to keep weeds from going to seed. After hand weeding, remove the weeds from the property. Don't let them go to seed and infest the adjacent pots.

Also, recycled potting media tends to be loaded with weeds. Use recycled media for potting woody ornamentals in which you can use broad-spectrum herbicides.

Preemergence Herbicides

Perennials are sensitive to many of the common nursery herbicides, particularly those that control a broad spectrum of broadleaf weeds. Scotts Ornamental Herbicide 2, Rout, Regal O-O, and Ronstar all control most weeds in container nurseries but have been shown to injure many perennials. Less efficacious herbicides are more likely to be safe on herbaceous crops but will, of course, not control as broad a spectrum of weeds.

To choose the best herbicide for your nursery, there are two basic strategies to consider: KISS and optimized.

In the KISS (Keep It Simple, Silly) method, you either rely exclusively on hand weeding or you choose a marginally effective preemergence herbicide that's safe on the majority of perennials being grown and supplement that with frequent hand weeding. Research has shown that even a marginally effective herbicide can be cost-effective, reducing the time required for hand weeding.

In an optimized weed control program, you choose the most effective preemergence herbicide labeled for each species. This option requires much more planning and on-site experimentation, but it provides the best, most cost-effective weed control. (For information on herbicides for specific crops, consult the chart below and the sources at the end of this article.)

Due to the tremendous diversity of plants produced in perennial nurseries, the KISS method seems to be the prudent choice for most growers. Larger nurseries, growing greater numbers of each species, may benefit from the optimized approach.

There are some species of herbaceous ornamentals that are even sensitive to our "safest" herbicides. If the herbicide isn't labeled for use on the species you're growing, run small trials before you treat the entire inventory. Also, remember that very young plants are more sensitive to herbicides than older plants, so liner producers will have to rely primarily on sanitation, while finished plant producers can use herbicides more widely. After potting liners,

Trade Name	Active Ingredient
Barricade 65 DG or Regalkade 0.5G	**prodiamine**
Fairly broad spectrum of weed control. Safe on many herbaceous ornamentals. Granule is much safer than spray.	
Corral or Pendulum 2G	**pendimethalin**
Fairly broad spectrum of weed control, including annual grasses, spurge, chickweed, and others. Granular formation is much safer than spray. Safe on many herbaceous ornamentals, but injures some species with foliage that traps granules.	
Devrinol 2G or 5G	**napropamide**
Somewhat narrow spectrum of weeds controlled in containers—primarily annual grasses. Safe on many herbaceous ornamentals but not widely tested on perennials.	
Dimension 1EC or 0.25G	**dithiopyr**
Fairly broad spectrum of weed control when used at the highest labeled rate. Primarily used for crabgrass control in turf but recent label expansion included many herbaceous ornamentals.	
Snapshot TG	**isoxaben and trifluralin**
Broader weed-control spectrum than the others listed here, but can severely injure several herbaceous perennials (most notably foxglove and most annual bedding plants).	
Surflan, XL	**oryzalin, oryzalin and benefin**
Broad-spectrum weed control. Safe on several "blue-collar perennials" such as hostas, astilbes, daylilies, and irises, but the most injurious of the herbicides listed here on many herbaceous perennials. The granular formulation (XL) is much safer than the spray.	
Treflan 5G or Preen	**trifluralin**
Controls annual grasses and a few broadleaf weeds. The weakest weed control of the herbicides is listed here, but also the safest herbicide on herbaceous ornamentals. The only species I have injured with Treflan 5G is *Lantana,* which had reduced flower counts but recovered.	

irrigate to settle the soil. The next day, apply the preemergence herbicide to dry foliage and irrigate to incorporate the herbicide into the media and wash it from the foliage.

Remember: No herbicide will control all weeds. Supplement with frequent hand weeding to reduce spread and secondary infestations. Effective herbicide programs will be more effective when combined with diligent sanitation.

Resources

Weed Control Suggestions for Christmas Trees, Woody Ornamentals, and Flowers, by Skroch, Neal, Derr, and Senesac. AG-427. To order: Send $7.50 (this includes shipping and handling) to Publications, NCSU, Box 7603, Raleigh, NC 27695-7603.

Weed Management Guide for Herbaceous Ornamentals, by Andrew Senesac. WeedFacts #1. To order: Send $1.25 to WeedFacts, Department of Floriculture and Ornamental Horticulture, 20 Plant Science Bldg., Cornell University, Ithaca, NY 14850.

Pest Management, June 2000.

The Weed-Free Greenhouse

You've heard of IPM—integrated pest management. How about IWM— integrated weed management? Tina Smith, an extension educator with the University of Massachusetts in Amherst, writes about IWM in UMass's *Floral Notes.*

Like IPM, IWM uses cultural, nonchemical, and selective chemical control of weeds.

Cultural controls include hand weeding, physical barriers (ground cloth), and solarization (allowing the weeds to dry up in an empty greenhouse).

Tina suggests prevention as the best way to manage weeds. Where you don't have paving, use weed-block fabric. Don't cover it with mulch or gravel, or you'll create another place for weeds to grow. Manually pull weeds along edges and in cracks, or use a herbicide selectively.

About herbicides: Few are labeled for use in the greenhouse due to the potential for crop injury or death. Injury happens in a number of ways, including spray drift because of fans and volatilization (the herbicide changing from a liquid to a gas). Herbicide vapors can easily build up within an enclosed greenhouse and injure susceptible plants.

Herbicides are classified according to the stage of weed growth affected. Preemergence herbicides are applied before weeds emerge, and they provide residual control of weed seedlings. Currently, no preemergence herbicides are labeled for greenhouse use. Surflan (oryzalin) is no longer registered for use in enclosed greenhouses.

In the greenhouse, several postemergence herbicides can be used under benches and on the floors. Irrigating crops too soon after applying an herbicide can wash it off and reduce its effectiveness.

Below is a list of herbicides currently labeled for use in greenhouses, and their manufacturers. If any information in the list is inconsistent with the label, always follow the label instructions.

Managing weeds outside the greenhouse is important to eliminate a major source of airborne weed seed and to prevent perennial weeds from growing under the foundation and into the greenhouse. Maintain a ten- to twenty- foot weed-free barrier around your greenhouses, using weed-block fabric or postemergent and soil residue herbicides. Surflan, Surflan combined with Reward, Finale, or Roundup can be used for post- and preemergent weed control. While spraying any weeds around the greenhouse with any

Trade Name	Common Name	Use with Crop in House?	Mode of Action/ Target Weeds	REI
Envoy (Valent)	Clethodim	Yes	Selective, contact/ Annual and perennial grasses	12 hours
Finale (AgrEvo)	Glufosinate-ammonium	Yes	Nonselective, systemic/ Annual and perennial grasses, broadleaf weeds	12 hours
Reward (Zeneca)	Diquat dibromide	Yes	Nonselective, contact/ Annual weeds	24 hours
Scythe (Mycogen)	Pelargonic acid	Yes	Nonselective, contact/ Annual and perennial broadleaf and grass weeds; most mosses and cryptogams	24 hours
Roundup DryPak (Monsanto)	Glyphosate	No	Nonselective, systemic/ Annual grasses and broadleaf weeds	12 hours
Roundup Pro (Monsanto)	Glyphosate	No	Nonselective, systemic/ Annual and perennial weeds	4 hours

herbicide, close windows and vents to prevent spray drift from entering. Also, use knockdown insecticide first to kill any flying insects that might leave the weeds and enter the greenhouse.

Culture Notes, September 1999.

Using Resistant Varieties for Chrysanthemum Pest Management

Kevin M. Heinz and Steve Thompson

Many growers choose a combination of several pest management tactics to achieve the best insect management on their crops; however, insecticides continue to be the principal approach growers use in building their management strategies. The problems associated with insecticide applications are well documented, problems that encourage the development and use of alternatives to traditional insecticides. One useful but often-overlooked alternative is resistant varieties.

Resistant varieties exhibit certain chemical, physiological, or morphological characteristics that interfere with insect pests' development or reproduction. In reality, few plants are completely immune to insect attack. However, plant varieties do vary in their degree of susceptibility. The genes for reduced susceptibility on plants may affect insect reproduction, survival, development time, or ability to digest and assimilate plant material. The products of these genes also may make the plant undesirable to insects, causing them to go elsewhere.

Selecting Resistant Varieties for Your Program

The several hundred commonly grown chrysanthemum varieties have various levels of resistance to leafminers, aphids, thrips, and other pests. Although you shouldn't overlook this when choosing chrysanthemum varieties to grow, your choice of insect resistance must be balanced with other desirable characteristics, such as growth habit and postharvest longevity.

How Can You Use Resistant Varieties?

If you choose resistant varieties for your pest management strategy, you can use them in several ways. One is to simply plant varieties that are resistant to a pest that's been a problem in the past. This requires you to keep accurate records from sampling data so you can anticipate problems. Identifying which pests are common problems is important because incorporating resistance to one species won't necessarily promote reduced susceptibility to others. A variety can be highly resistant to aphids, for example, but very susceptible to thrips or leafminers.

Another use for resistant varieties involves the way insects move through greenhouses. Flying insects will often settle on and "taste" a plant they encounter. Many insect problems originate outside the greenhouse, and if you put resistant chrysanthemum varieties near insect entry points, they'll be the first plants insects encounter. Less susceptible plants should cover a large enough area around points of entry so that they challenge all invading insects. This can discourage the development of large damaging pest populations. For this to be an effective strategy, you must know the main way that insects enter your greenhouse.

Interspersing resistant varieties with susceptible varieties in an "intercropping" arrangement may also reduce the rate at which a pest outbreak spreads, especially for slow-moving pests. Arrange different varieties in a regular alternating pattern when you plant so insect populations don't build up on large patches of susceptible plants. Also, you can alternate varieties with resistance to different pests if you need to manage more than one insect.

Don't rely on any single management tactic. Consider using less-susceptible varieties along with other practices to develop an overall management strategy.

Table 1 lists several chrysanthemum varieties and their relative level of susceptibility for three principal insect pests as well as white rust disease. The

Table 1. Chrysanthemum Resistance by Variety[a]

Variety	*Liriomyza* sp. Leafminers	*Frankliniella* sp. Thrips	Aphids	White rust	Color/form
Albert Heijn	High	High	Average	High	Pink/spider
Albert Heijn Royal	High			High	Dark Pink/spider
Albert Heijn White	High			High	White/spider
Anita	Average			Low	White/single
Casa	Average			Low	White/single
Casa Sunny	Average			Low	Yellow/single
Charm			Low		Pink/decorative
Diamond			Low		Yellow/decorative
Ellen van Langen	Low	High	Average	Low	White/single
Finmark	Average			High	White/anemone
Freedom	High			High	White/single
Hawaii	Low			High	Cream /decorative
Hawaii Yellow	Low			High	Yellow/decorative
Helsinki		Average	Average	High	White /semi-double
Improved Penny Lane	High	High	High		White/anemone
Iridon			Low		Yellow/decorative
Kes	Low	Low	Average	High	Pink/spider anemone
Klondike	High			Low	Red/single
Majesty	Low			High	Pink/single
Lineker	Low	Low	Above average	Low	Pink/single
Nikkei Improved	High			Low	White/spider
Palet		High	Low	Low	Pink/single
Paso Doble	High	Low	Low	High	White/anemone
Paso Doble Pink	High			High	Pink/anemone
Puma	High			Low	White/anemone
Puma Sunny	High			Low	Yellow/anemone
Pomona			Low		Pink/decorative
Reagan Improved	Low	Low	Average	Low	Pink/single
Reagan Red	Low			Low	Red/single
Reagan Orange	Low			Low	Orange/single
Reagan Salmon	Low			Low	Salmon/single
Reagan White	Low			Low	White/single
Sheena	Average	High	Average	Low	White/spider
Smile	Low			Low	White/spider
Stallion	Average	High	Above average	Average	White/mini spider
Statesman	High	High	High	High	Yellow/pompon
Symphony			Low		Pink/decorative
Tiger	Average	Low	High	High	Orange /single
Tigerrag	Average			High	Red/single
Touch	Low			Low	Yellow/anemone
Westland Regal	High			High	Violet /spider
Westland Winter	Average			High	White/spider
Westland Yellow	Average			High	Yellow/spider

[a]Relative level of resistance of selected chrysanthemum varieties to leafminers, aphids, thrips and white rust. A "high" ranking means little damage was observed, while a "low" ranking means substantial damage was observed. Blank spaces in the table indicate no data is available to assign a rank.

table demonstrates that chrysanthemum resistance doesn't provide control for all pests. Resistance to one pest group doesn't necessarily correlate well to resistance in another.

This isn't surprising when you look at insects' feeding behaviors: Leafminers feed on plant sap as adults, and larvae feed on internal plant tissues. Thrips pierce plant cells with their mouthparts to feed on cell contents. Aphids feed on plant sap running through plant phloem. White rust is a plant pathogen with different characteristics than insects possess. Hence, discovering a chrysanthemum variety resistant to all forms of attack is unlikely. However, the benefit of marketing desirable varieties is having many varieties with different flower colors and forms with insect or disease resistance.

Editor's note: The authors thank Nancy Rechigl, Yoder Bros. Inc., for her assistance and the American Floral Endowment for providing funding.

Pest Control, March 1997.

Highlights from SAF's Pest Conference

Margery Daughtrey and Christine Casey

The fourteenth annual Pest Management Conference, sponsored by the Society of American Florists, was held in Del Mar, California, February 21 to 23, 1998. More than two hundred growers, educators, and horticultural suppliers gathered to hear the very latest in pest management theory and practice. The meeting included lectures, workshops, and tours. Here are some highlights from the three-day event:

Virus Symptoms

Margery Daughtrey, Cornell University, began with a presentation on the symptoms of tospoviruses on flower crops. The "generic" symptoms that should trigger growers' suspicions include round brown, black, or white spots; necrosis (dead tissue) at the petiole end of the leaf; yellow mottling or variegation; death of young plants or the terminal foliage of older plants; stunting; brown or black stem cankers; vein necrosis; concentric ring spots or zonate spots; or even wilting or line patterns. Indicator plants are recommended for early detection of tospoviruses. Growers should confirm their suspicions with an in-house test kit or by sending samples to a laboratory.

Diane Ullman, University of California, Davis, also covered tospoviruses, which are vectored by thrips and cause significant losses for ornamentals producers. Her monitoring system uses indicator plants and directional sticky traps to locate sources of infective thrips before crop symptoms are seen. Removal of thrips sources can greatly reduce virus losses. (*Editor's note: See "A New Weapon to Fight INSV and TSWV" on p. xx for details of this research.*)

Leafminers

Michael Parrella, UC Davis, discussed the leafminer species that affect ornamentals, of which *Liriomyza trifolii* is the most important. Avid has been the standard for leafminer control since its introduction, but is now losing its effectiveness. A new material, Conserve (spinosad), gave 80 to 85% mortality in his greenhouse trials. Natural enemies are commercially available, but a prohibitively large number is needed for adequate control.

Whitefly

Mark Hoddle, UC Riverside, described how to use whitefly biological control. Parasitoid species, release rates, and combinations with insect growth regulators were evaluated in his studies. *Eretmocerus eremicus* was effective for silverleaf whitefly control and is best used in combination with Precision or Applaud. Enstar was not compatible with the natural enemies he tested. Biological control should be used only at low whitefly levels (less than one nymph per ten cuttings) and must be regularly evaluated.

Mildews

Mary Hausbeck, Michigan State University, covered downy and powdery mildew, stressing the difference between these two diseases. Symptoms of downy mildew are sometimes mistaken for spray injury or nutritional problems, so look for spores on leaf undersides with a hand lens when conditions are humid.

A new strobilurin material called Cygnus will soon be available for powdery mildew. In poinsettia powdery mildew trials, all cultivars tested were susceptible. Sprays of Systhane gave control on cuttings for thirty to fifty days after application to stock plants.

Rhizoctonia

Jan Hall, Paul Ecke Ranch, Encinitas, California, provided insights into *Rhizoctonia* management. Sanitation is critical, since the fungus survives in soil. Hardwood bark composts are naturally suppressive to *Rhizoctonia*. New biological fungicides for this disease include Deny, based on an antagonistic bacterium (soon be available as PathGuard), and a new *Gliocladium*-based product.

Methyl Bromide Alternatives

Jim MacDonald, UC Davis, discussed research into methyl bromide alternatives. In *Fusarium* studies, methyl iodide was shown to be more effective than methyl bromide, while Basamid only reached a one-foot soil depth. The use of three-foot steel rods to deliver electrical current to the soil eliminated *Fusarium* to a two-foot depth. Use of a plastic barrier over a carnation bed protected against *Fusarium* recontamination. Antagonistic microbes have a role in preventing reinfestation after pasteurization, as well.

Elizabeth Mitcham, UC Davis, discussed her research into the utility of controlled atmospheres (less than 0.5% oxygen and 20 to 95% carbon dioxide) and other alternatives to methyl bromide for postharvest pest control. Initial results with two-spotted spider mite and western flower thrips are promising. Investigations with leafminer, melon aphid, and whitefly will begin soon.

Fungus Gnats and Shore Flies

Dick Lindquist, Ohio State University, described how fungus gnats and shore flies are more than just nuisance pests, as they can spread pathogens such as *Pythium* and *Thielaviopsis*. Fungus gnat larvae also injure cuttings. His media trials showed the highest number of fungus gnat larvae in 100% coir, while other studies demonstrated the efficacy of fungus gnat biological control. Distance, a new growth regulator, gave good control of shore fly adults compared to DuraGuard.

Fusarium

Ann Chase, Chase Research Gardens, Mount Aukum, California, covered *Fusarium* management. Strategies include using nitrate rather than ammonium nitrogen, and adding lime to the mix. Heavily fertilized crops are more susceptible to *Fusarium*. Trials of fungicides for control of *Fusarium* wilt of cyclamen have shown incomplete control with Heritage, Medallion, or Terraguard. A new *Fusarium* crown rot on lisianthus is under study in Florida. Medallion and Chipco 26019 plus Domain have shown the best results thus far.

Resistance Management

Karen Robb, UC Cooperative Extension, San Diego, stated that pesticide resistance had been documented in 504 insect species. Resistance management strategies include using IPM tactics to limit pesticide use, rotation of pesticide classes, using less persistent materials, and avoiding tank mixes of chemicals in the same class.

Margery Daughtrey discussed resistance development in greenhouse fungal pathogens. No cases of *Botrytis* fungicide control failure have been documented in North America, although resistant *Botrytis* has been detected in Canada, Pennsylvania, Connecticut, and South Carolina. Metalaxyl-resistant *Pythium* isolates have also been found in samples from a number of states. Rotation of active ingredients and use of IPM strategies for disease management is the best recourse.

Botrytis

Mary Hausbeck described the attention to sanitation required for effective *Botrytis* management. It sporulates on dead tissues, so trash containers should always be covered. Spores are naturally released into the greenhouse mid-morning and mid-afternoon, and worker activity also causes spore release. Forcing heated air under benches helps to reduce the relative humidity within the plant canopy. A white plastic mulch over stock plant pots also helps. Greenhouse plastic coverings that block UV light transmission reduce *Botrytis* reproduction. Apply a protectant fungicide before cleaning plants of diseased leaves and blossoms. The relative humidity should be lowered to less than 65% for three days after cleaning.

Snails and Slugs

Michael Parrella discussed snails and slugs, which are quarantined by several states. They prefer cool, moist areas, although many snails are tolerant of warm, dry situations and can survive for extended periods until conditions become more favorable. Cultural controls are effective and should be used in rotation with metaldehyde, the only available chemical control. Natural enemies such as birds and predatory beetles also play a role in control of snails and slugs.

Aphids

Dick Lindquist discussed green peach and melon aphids, the most common aphid pests in greenhouses. New products are under development for aphid control. Relay is a systemic that is compatible with biological controls; Conquest is similar to Marathon. Cinnacure is currently registered, but not widely available. While there are many aphid natural enemies, they're often not compatible with controls used for other pests.

Pythium and *Phytophthora*

Ann Chase described the various impacts that *Pythium* and *Phytophthora* can have on plants. In her studies, she found metalaxyl-resistant isolates for several different *Pythium* species. This may represent preexisting variability within the species, rather than the result of overuse of metalaxyl. The biofungicides RootShield and Gliogard were shown to protect plants against *Pythium,* but overdoses may cause phytotoxicity.

Water Treatments

Jim MacDonald presented studies on the treatment of recycled irrigation water to eliminate plant pathogens. *Pythium* and *Phytophthora* may be reintroduced to filtered water. UV radiation from pulsed laser and pulsed xenon sources were effective at treating recycled irrigation water. A ten-minute exposure to sodium hypochlorite was required to eliminate *Fusarium* spores, while ozonation gave nearly 100% mortality after twelve minutes. Ozone has less potential for phytotoxicity than chlorine.

April 1998.

Chapter 5
General Pest Control

Greenhouse Pest Control: Then and Now

Richard Lindquist

What were the most important insect and mite pest problems on greenhouse ornamentals since 1930? In 1930, the major insect and mite pest problems included chrysanthemum and rose midges, fungus gnat, leafminer, aphid, thrips, whitefly, scale, mealybug, plant bug, root weevils (e.g. black vine weevil), caterpillar, bulb mite, spider mite, and cyclamen mite.

High-volume spraying well before 1930. The same basic methods are still used, although protective clothing requirements have changed.

Today, chrysanthemum midge isn't mentioned as a problem in greenhouses. Rose midge only occasionally damages greenhouse roses. Unfortunately, most of these other pests are still around and are still causing problems on a wide range of crops. In many cases, different species than those found in 1930 are attacking today. Some of this may be due to renaming the same insect, some to different species moving into the picture. For example, thrips—but not western flower thrips—was a problem in earlier decades. The "other" thrips are still around, but the predominant pest is now the western flower thrips. Greenhouse whitefly was a widespread pest on greenhouse crops sixty and seventy years ago, as it is now. Sweetpotato

127

and silverleaf whiteflies weren't mentioned. These whiteflies have entered the greenhouse picture only recently. Shore fly wasn't mentioned in earlier pest control publications. Leafminer has come and gone (and come again). So, not much has changed in terms of basic pest problems. A grower from 1930 would feel comfortable with today's pest problems.

Control

What about approaches to pest management? Again, there isn't much difference in what was suggested for an effective pest management program and what we use today.

- Recognize the pest or pest injury.
- Know the basics of the pests' life cycles.
- Maintain a vegetation-free area surrounding greenhouses.
- Install screening over doors and ventilators.
- Sanitation—remove weeds; destroy crop residue.
- Use (or at least encourage) biological controls.
- Use pesticides but only when necessary.

It's difficult to argue against any one thing on the above list. However, there are some significant differences between then and now. We know a great deal more about insect and mite biology and detection—and learn more each year. Damage thresholds are being developed for some crops. Many companies rear and/or distribute biological controls to commercial growers and interior landscapers. Pesticides have changed as well.

Aerosol application in a greenhouse tomato crop. Note the protective headgear—one could be stylish and still apply pesticides.

Probably the biggest change is that we aren't thinking about pest management and disease management as separate things but also as parts of integrated crop production systems. In other words, what's done in one aspect of crop production may and probably does also affect other aspects. Also changing is the idea that insect and mite pest management must be completed by the time crops are mature. Postproduction pest management is an area of research and development with much promise.

Pesticides

Then, as now, pesticides are still the primary means of pest management. The table includes many pesticides recommended for controlling major insect and mite pests from 1930 to present day. The 1930s were dominated by pesticides containing arsenic and cyanide—not very benign stuff. The 1960s (back in my day) saw the use of many of the chlorinated hydrocarbon and organophosphate compounds— many very toxic. In the 1980s carbamate pesticides (some very toxic) were used, along with some pyrethroids. Although some old-timers such as pyrethrum, oil, soap, and nicotine are still used today, other products such as Paris green, lead arsenate, cyanide, mercuric chloride, DDT, parathion, demeton (Systox), and aldicarb (Temik) are no longer used.

We now have insect growth regulators, new conventional chemical classes such as imidacloprid (Marathon) and pyridaben (San-mite), and microbials such as *Beauveria bassiana* (BotaniGard, Naturalis-O). Although the goals of pesticide application remain the same, the products used are different.

Application methods in 1930 were quite crude by today's standards, but high-volume sprays were used much as they are now. Alternative application methods have also been used for decades.

Pesticides Recommended for Greenhouse Ornamentals from 1930 to Present Day

1930s

nicotine • cyanide • paris green + sugar • lead arsenate • white arsenic • mercuric chloride • pyrethrum + soap • oil • soap • sulfur dust (mites only)

1960s

nicotine • cyanide • DDT • lindane • endosulfan • demeton • disulfoton • dimethoate • naled • oxydemeton-methyl • dichlorvos • diazinon • malathion • parathion • sulfotepp • carbaryl • pyrethrum • oil • soap • dicofol • tetradifon (mites only)

1980s

nicotine • endosulfan • methoxychlor • acephate • disulfoton • dimethoate • oxydemeton-methyl • dichlorvos • naled • diazinon • malathion • sulfotepp • aldicarb • carbaryl • pirimicarb • methomyl • oxamyl • fluvalinate • permethrin • resmethrin • phenothrin • oil • soap • kinoprene • *Bacillus thuringiensis* • dicofol • dienochlor • cyhexatin • hexakis (mites only)

1990s

nicotine • endosulfan • lindane • acephate • chlorpyrifos • diazinon • dichlorvos • malathion • naled • sulfotepp • bendiocarb • methiocarb • fluvalinate • bifenthrin • cyfluthrin • fenpropathrin • permethrin • resmethrin • phenothrin • lamba-cyhalothrin • kinoprene II • cyromaine • diflubenzuron • azadirachtin • fenoxycarb • abamectin • imidacloprid • pyridaben • oil • soap • *Beauveria bassiana* • *Bacillus thuringiensis* • dicofol (mites only) • dienochlor

Dusts were used more in the past than they are now. Self-contained aerosol "bombs" containing pesticides such as parathion were widely used in the 1950s and 1960s. Smoke generators have also been available for decades. Low-volume sprays aren't new. Thermal fog applicators have been used for at least thirty-five years. We now have other application equipment such as electrostatic sprayers that improve coverage and deposition. Regardless, the goals of pesticide application remain the same: to get the correct pesticide to the correct place in the correct amount at the correct time(s) to control pests.

Nonchemical Control

Various biological and physical methods have been used for insect and mite control. Some of these were quite interesting but probably aren't practical today. Others were practical then and are still effective today.

Biological controls on greenhouse ornamentals were encouraged in the 1930s and again in the late 1980s and 1990s. Then, as now, success has been variable. In the middle of this period, many people assumed pesticides were the only answers to insect and mite problems—and they were probably used to excess—so there was little discussion of biological or other nonchemical controls. However, during this time biological controls became common on greenhouse vegetable crops in many parts of the world, and they successfully controlled many of the same pests that occur on ornamentals.

Physical controls were common in 1930. Injecting plants with water to reduce populations of spider mites and thrips, for example, is one control (by the way, this is still effective—except for encouraging the development of some plant pathogens if foliage remains wet at night). Soaking bulbs in hot water was suggested as a control for bulb mites. This worked then and still works. Hot water is now used for controlling insects in some cut flowers after harvest. Screening to exclude certain pests was suggested in 1930 and is used in many parts of the U.S. now.

Sometimes physical control included physically removing insects by hand. A 1930s recommendation for controlling adults (moths) of an insect called the greenhouse leaf tier (caterpillars feed on and tie leaves together) is as follows: "A variation of hand picking, as recommended by the Ohio Experiment Station, consists of driving the moths into the flame of a gasoline torch. This is accomplished by carrying the torch slowly up and down the greenhouse walks and gently striking the plants with a fly swatter or stick to disturb the moths, which will then fly into the flame, burn off their wings and drop to the ground. An assistant should follow the one carrying the torch to kill these moths by stepping on them as fast as they fall." This method was also suggested for controlling moths of other caterpillar species. Yet another variation consisted of hitting the moths with a fly swatter as they flew into the glass walls at dusk.

Actually, the pest disturbance method may not be obsolete. I have often thought that a variation of the flaming torch method—replacing the flaming torch with sticky traps—could be used to reduce shore fly numbers on some bedding plants—if it were automated and done repeatedly.

That's a brief summary of the history of insect and mite management on greenhouse ornamental plants. The basic pests are the same; the basic approaches to control are the same—we just have some modified tools and more knowledge (hopefully better) with which to accomplish the tasks. As long as plants are continuously produced in nice, warm, confined spaces, this situation should continue. Job security is wonderful.

Pest Control, May 1997.

New Biorational Materials for Insect Control

Dr. Roger C. Styer

At the Society of American Florists' thirteenth Conference on Insect and Disease Management held in February in Orlando, Florida, Dr. Michael Parrella, University of California, Davis, provided an update on his research with biorational pesticides.

Biorationals include many types of new products: natural products (metabolites), synthesized analogues of naturally occurring biochemicals, microbial control agents, or inorganic products. The great news is that these reduced-risk chemicals are getting reviewed and approved faster, even in California, due to the Food Quality Protection Act. You'll need to incorporate them into an IPM program, as none are as good as broad-spectrum chemicals and they don't work as well alone. Following is a rundown of several with great promise:

BotaniGard is a new product from Mycotech Corporation, Butte, Montana, and is the naturally occurring fungus *Beauveria bassiana*. This product should be very similar to Naturalis-O but may be a different strain or a different formulation. It's active against a wide range of greenhouse pests, particularly whiteflies, thrips, and aphids. Spores land on the insect cuticle, enter the gut, multiply, then kill the insect. You'll probably need to use this product regularly for best action. Spores appear to spread on infected adults. The fungus works better for thrips control when sprayed into flowers.

Cinnacure, a new material not yet released, is a synthetic analogue of naturally occurring cinnamon oil with good activity against spider mites, melon aphids, thrips, and powdery mildew. It's similar to insecticidal soap, in that the material needs to be in contact with the pest at the time of application. A rate of 0.3% and a proprietary formulation have been identified. This should avoid phytotoxicity problems that showed up in earlier research.

Spinosad is produced through fermentation of a naturally occurring material produced by the fungus *Saccharopolyspora spinosa*. Success, the trade name for this product from Dow Elanco, has broad activity against a wide range of pests including worms, leafminers, and thrips. This material's mode of action is similar to Marathon. It can work as well as Avid and Citation and could be incorporated into a pesticide rotation for resistance management.

Michael is also working on some interesting insect growth regulators. Insect growth regulator categories include juvenile hormone products, which give a slower kill, taking up to twenty-one days to act, and chitin synthesis inhibitors, which kill quickly, within seven days. Examples of juvenile hormone products include Enstar II, Precision, and Preclude; chitin synthesis inhibitors include Citation and Adept.

Michael has researched Nylar (pyriproxyfen) and Applaud (buprofezin) to examine their control of whiteflies on poinsettias. Nylar is a juvenile hormone product that provides excellent control but takes 21 days to achieve it. Applaud is a chitin synthesis inhibitor that provides excellent control in just seven days. These materials may be able to take some of the pressure off of the extensively used Marathon.

GrowerTalks in Brief, May 1997.

Insect-Killing Fungi: Floriculture's IPM Future?

Brook C. Murphy, Tunyalee Morisawa, and Michael P. Parrella

Faced with increasing insect resistance, spiraling costs and the continued loss of product registrations, many growers are searching for alternatives to their standard pesticide programs. Unfortunately, cost-effective alternatives to

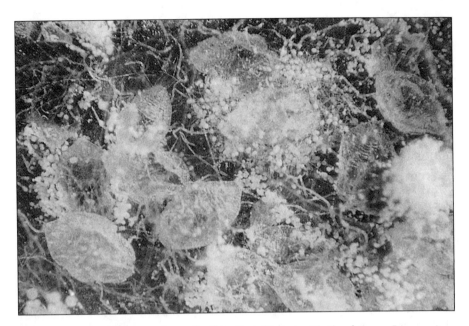

Immature silverleaf whitefly on poinsettia infected with the fungus *Beauveria bassiana.*
Photo by Michael P. Parrella.

conventional pesticides that control key greenhouse pests without the side effects of traditional pesticides have been difficult to find.

Many biological controls, while important in IPM, can't be used where pesticides are regularly applied. However, with recent changes in registration procedures for reduced-risk pesticides, the development of "soft" pesticides such as microbial and natural biochemical pesticides has been advancing rapidly. One potential candidate that has received federal registration for use on ornamental crops is a microbial insecticide containing the entomopathogenic fungus *Beauveria bassiana*. This fungus is a naturally occurring insect pathogen and is an important mortality agent of many arthropods. Two commercial products containing the fungus are available. BotaniGard is formulated as a wettable powder (WP) or emulsifiable oil (ES) and is produced by Mycotech Corporation, Butte, Montana; Naturalis-O is produced by Troy Biosciences Inc., Lake Placid, Florida. All of our work at the University of California, Davis with this fungus has been with the BotaniGard WP and ES formulations. In both commercial and experimental trials on poinsettias, we evaluated BotaniGard's effectiveness for whitefly control.

How the Fungus Works

Beauveria bassiana spores are formulated to mix readily in water and are applied using standard spray equipment. The fungus kills insects either by direct application from spray equipment or through secondary contact with spores on plant foliage. When spores come into contact with an acceptable host, a germ tube penetrates the insect's cuticle and feeds from the host body, resulting in the death of the host. In most cases, it takes eight to ten fungal spores on an insect to cause fungal infection and the subsequent death of the insect.

One drawback with most fungi is the high relative humidity and temperature needed for successful germination. However, *B. bassiana* seems to be capable of infecting insect pests in a wide range of environmental conditions. The warm temperatures and relatively high humidity in greenhouses are ideal environments for using this fungal pathogen. Because fungal spores kill insects through direct contact, good spray coverage is essential for achieving adequate control.

Beauveria bassiana fungi have demonstrated a number of attributes that make them potentially ideal reduced-risk insecticides for controlling major greenhouse pests: 1) they're relatively host specific; 2) they have low mammalian toxicity; 3) they can be cheaply mass-produced on artificial media; 4) they can be effective over a wide range of environmental conditions; and 5) they can kill a high proportion of the target population. In addition, BotaniGard WP can be readily applied using standard spray equipment and can be tank mixed with a number of conventional pesticides and even a few fungicides.

Commercial and Experimental Trials

Though we've conducted evaluations using BotaniGard against thrips, aphids, and whiteflies, only whiteflies will be discussed in this article. We used the sweetpotato whitefly (SWF) attacking poinsettias as a model system for evaluating BotaniGard as an alternative to conventional pesticides for whitefly control in greenhouses.

We conducted evaluations of BotaniGard's commercial performance under both experimental and commercial greenhouse conditions. The experimental greenhouse trials were conducted to evaluate BotaniGard's ability to rescue the poinsettia crop from excessively high SWF numbers; to evaluate the compatibility of using BotaniGard with a key whitefly parasitoid, *Encarsia formosa*; and to identify the minimum effective dosage of

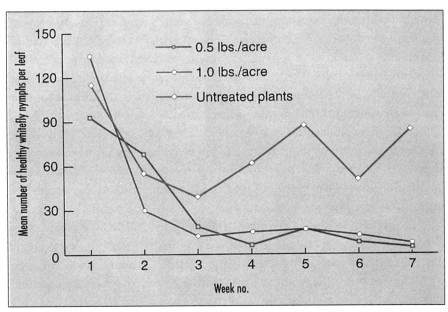

Figure 1. Mean number of sweetpotato whitefly nymphs per leaf for untreated plants and 0.5-lb. BotaniGard and 1.0-lb. BotaniGard treatments.

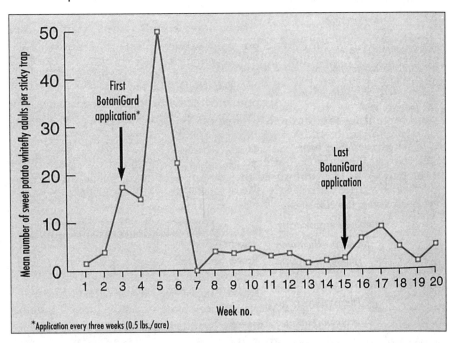

Figure 2. Mean number of sweetpotato whitefly adults caught on sticky traps in potted poinsettias treated with a 0.5-lb. BotaniGard treatment.

BotaniGard needed for economic control of whitefly populations. Then, with the knowledge gained from the experimental trials, we conducted a trial on commercial poinsettias to validate the results from the experimental trials.

For the experimental trials, we used sealed cages containing five potted poinsettias ('Lilo Red') infested with SWF at approximately 100 nymphs per leaf. We evaluated three treatments: an untreated control, BotaniGard WP at 0.5 lbs./acre equivalent and BotaniGard WP at 1.0 lbs./acre equivalent. We made six applications at seven-day intervals. Each week, we assessed the average density of SWF nymphs (fig. 1), the proportion of the nymphal populations infected with the fungus and the proportion of the population parasitized by *E. formosa* parasites.

For the commercial greenhouse trial, we used a 15,000-sq. ft. potted poinsettia range with low numbers of SWF present. We used a single rate of BotaniGard WP at 0.5 lbs./acre for pest control. Because the low numbers of SWF nymphs on leaves made monitoring nymphal densities unfeasible, we set out 12 yellow sticky traps and counted weekly to assess relative SWF numbers. The trial ran for 20 weeks from July through December (fig. 2).

Results

In the experimental greenhouse trials, one week after the first application (week 2) the rate of fungal infection for SWF nymphs began to rise sharply. We recorded nymphal infection rates of 30% for the 0.5-lb. rate and 50% for the 1.0-lb. rate after seven days. For the remaining five weeks, infection rates varied, but they averaged approximately 50% for both treatments. We found no fungal infections in SWF nymphs on untreated plants. The results revealed that the *B. bassiana* fungus was infecting a large proportion of the SWF nymph population. Infection by the fungus always results in the death of the pest. Also, although the higher BotaniGard dosage achieved a greater initial infection rate, both dosages infected nymphs similarly when averaged over time.

SWF nymph densities on leaves also declined rapidly as infection rates increased. Reductions of 54 and 70% were achieved for the 0.5- and 1.0-lb. rates respectively, after the second application (week 3). For the remaining four weeks, SWF densities remained low, ranging from four to 15 nymphs per leaf. SWF on untreated plants remained significantly higher, ranging from 40 to 90 nymphs per leaf (fig. 1). Results indicated that under intense SWF pressure, BotaniGard WP could significantly reduce whitefly numbers and potentially rescue a crop that could otherwise be unmarketable.

The presence of *Encarsia formosa* during these trials allowed us to evaluate the compatibility of BotaniGard WP with a key natural enemy of SWF nymphs. Comparing the nymphal parasitism rates by *E. formosa* between the two fungal treatments and the untreated control plants revealed no significant differences in the proportion parasitized. This indicates that the fungus impacts the target pest population to a much greater degree than do natural enemies.

For the commercial trial, we began BotaniGard WP applications as the adult SWF population began to increase as indicated by the trap captures. Thereafter, we made applications approximately every three weeks for the remainder of the cropping cycle (five applications total). After the first spray, adult captures initially began to increase, then declined and remained low (less than 10 SWF adults per trap per week) for the remainder of the season (fig. 2). The trial revealed that by initiating applications early (before SWF populations threatened the crop) the 0.5-lb. rate was sufficient to maintain SWF populations below damaging levels. We produced an entire crop of commercial poinsettias using BotaniGard WP as the only pest control material.

The results of our trials evaluating BotaniGard indicate that a viable alternative to conventional pesticides may finally be available to the greenhouse grower. It's safe for workers to use in enclosed greenhouse environments, doesn't pose a threat to the environment, and is compatible with natural enemies. Furthermore, it's versatile enough to apply as a conventional pesticide and even to tank mix with other pesticides. In addition, its compatibility with natural enemies truly makes this an IPM material and may herald the use of biological control for the ornamental industry.

December 1997.

Insect Growth Regulators

Michael P. Parrella and Brook C. Murphy

One group of reduced risk pesticides that has seen considerable development activity in the 1990s is the insect growth regulators (IGRs). Many large agrochemical companies are developing or have received registration for IGR compounds in major agrochemical markets; pesticides with a narrow spectrum of activity (i.e., controlling one or more important pests) are often recommended in control programs, and growers' use and understanding of IPM has expanded considerably.

While there is excitement about the availability of IGRs in the U.S., two factors remain poorly understood: the stage specificity of these products and their compatibility with natural enemies. This information is critical to the proper use of IGRs in IPM programs. As such, for the past several years, with funding from the American Floral Endowment, the California Cut Flower Commission, and the California Association of Nurserymen, we've been studying the stage specificity of several IGRs in control evaluations using two key greenhouse pests: the silverleaf whitefly and fungus gnat.

Types of IGRs

IGRs fall into several categories, but the two most important in terms of commercial viability are juvenile hormone analogs and chitin synthesis inhibitors. Juvenile hormone analogs mimic an important metabolic hormone unique to insects that affects growth and development. Typically, the effects of juvenile hormone mimics may not be seen immediately—the impact may occur when the insect is ready to pupate. At this time in an insect's life, the ratio between the juvenile hormone and other hormones in the insect is critical. By disrupting this ratio, the insect may be unable to molt into the pupal stage or becomes deformed and is unable to complete development into an adult.

These materials were first developed for mosquito control because immature mosquitoes were of no consequence. As long as they could be kept from developing into adults, successful control was achieved. The same scenario for pest control on ornamental plants is more difficult to rationalize because immature stages of whiteflies and other pests are very damaging. Nonetheless, this is an important point to keep in mind when using IGRs of the juvenile hormone type—they most likely won't act very quickly.

Because juvenile hormone is also important for the maturation of adult sexual characteristics, adults that emerge from treated pupae or that contact fresh residues after emergence may be unable to lay viable eggs. Chitin synthesis inhibitors act by disrupting the formation of chitin, a component of the insect cuticle (skin). Without the proper formation of the cuticle, the insect will die. Because the insect sheds and reforms its cuticle with each molt, chitin synthesis inhibitors may act at any time during an insect's immature development. They generally act more quickly than juvenile hormone mimics do. In general, both classes of IGRs have little or no impact on adults; they're primarily active on immature stages of the targeted pest.

Materials we evaluated included juvenile hormone mimics Enstar II (kinoprene 10E, registered by Sandoz), Precision (phenoxycarb 25W, registered by Novartis), and pyriproxifen 0.83E (registration pending by Valent); and the chitin synthesis inhibitors Adept (diflubenzuron 25W, registered by Uniroyal), Trigard (cyromazine 75W, registered by Novartis, registration pending in California), and Applaud (buprofezin 40SC, registered on field crops by AgrEvo). All materials were used at recommended label rates.

Performance Trials against Whiteflies

Silverleaf whitefly colonies were maintained on poinsettia plants (cultivar 'Red Sails') in a greenhouse at the University of California, Davis. Rooted poinsettia cuttings, obtained from a commercial producer, were individually potted into four-inch pots and kept in a greenhouse under normal production practices. When plants were four to six weeks old, they were trimmed back to three leaves and exposed to whiteflies in large colony cages. Exposure times varied with the size of the whitefly population but generally were between four and twenty hours. After removal, plants were placed in isolated greenhouses free of whiteflies and held at approximately 86°F. With this protocol, we were able to evaluate the performance of selected IGRs on specific life stages of whitefly.

Sprays were applied to runoff using a small backpack sprayer at 60 psi; only one application was made to each whitefly stage category. By making applications to plants a few days after exposure to adult whiteflies, mortality to the egg stage could be evaluated. By spraying seven days after exposure, mortality of young instars could be determined. Following this protocol, spraying fourteen and twenty-one days after exposure targeted third and fourth instars and red-eyed pupae (late fourth instars), respectively. Pretreatment counts of whiteflies on plants were made the day before IGR application. Application dates are approximate, which allowed for stage-specific timing, despite variation in cohort development due to variation in temperature, whitefly density, and host plant quality.

Four IGRs—kinoprene, phenoxycarb, buprofezin, and pyriproxifen—were evaluated against the various life stages; an untreated control was also included. Post-treatment counts (the number of individual whiteflies surviving) on all plants were made seven, fourteen, and, twenty-one days after treatment, depending on the stage treated. For red-eyed pupae, recording of survivors (in some cases this would include counting empty exuviae) was done only seven days after treatment. In this way we followed a cohort of similar age through development to the adult stage after one application of each IGR. Ten plants were used for each IGR/whitefly stage category. After application, plants were placed in a randomized design in an isolated and clean greenhouse and maintained under normal poinsettia production practices.

Results against Whiteflies

When IGRs were applied to the egg stage, only buprofezin provided control that was significantly different from the control, but this was observed twenty-one days after application. Other IGRs didn't have an impact at any time when compared to the control. Against the early instars, buprofezin had a more immediate effect and provided significant control seven days after treatment (table 1). When this cohort was followed through time, the effects of the other IGRs (with the exception of phenoxycarb) became apparent as whiteflies entered the late fourth instar or emerged as adults. Twenty-one days after treatment, pyriproxifen provided control equivalent to buprofezin, while kinoprene significantly reduced populations compared to the control. No significant effects were observed when any IGRs were applied to third and fourth instars or red-eyed pupae even when these were followed through to adult emergence. In a separate series of experiments not described here, no IGR affected the fertility of adult whiteflies treated as immatures.

Table 1.	**IGR Performance against Early Developmental Stages of Silverleaf Whitefly**				
Material	**Number of plants**	**Pretreatment count Number per plant**	**Post-treatment % control**		
			7 days	**14 days**	**21 days**
Control	10	94.4	0.04	18.2	22.7
Buprofezin	10	175.2	99.0	99.2	99.3
Kinoprene	10	159.1	45.4	52.3	86.9
Pyriproxifen	10	105.1	0.04	15.0	99.6
Fenoxycarb	10	168.7	9.5	24.4	61.6

Plants were exposed to whitefly colonies for four hours; applications were made six to eight days later. Recommended label rates of all products were used.

Performance against Fungus Gnats

Quarter-inch-thick potato slices were placed on the top of soil in four-inch pots and buried so they were flush with the soil surface. Pots were then placed under greenhouse benches where heavy populations of fungus gnats (*Bradysia* spp.) resided. Pots were left for twenty-four hours to allow for fungus gnat oviposition on the potato slices; they were then placed on benches in the same greenhouse. Pots were placed in plastic flats and fitted with an organdy cover supported with polystyrene tubing. The purpose of the screening was to prevent further oviposition by adult fungus gnats.

After five days, all larvae on and adjacent to the potato slices were counted, then drenched (75 ml of water per pot) with the following IGRs: dimilin, phenoxycarb, and cyromazine. An untreated control was also included. Post-treatment counts of larvae were made seven and fourteen days after treatment. A yellow sticky card hung within the organdy cages above the pots recorded adult emergence from the pots twenty-one days after treatment. There were four flats per pesticide treatment with five pots per flat.

Results against Fungus Gnats

A reduction in fungus gnat larvae was noted seven days after treatment (table 2), with diflubenzuron and cyromazine providing a significant reduction compared to the control. After fourteen days, fungus gnat larvae began to pupate, so no differences were observed comparing all treatments to the control. However, all materials significantly reduced adult emergence compared to the control twenty-one days after treatment.

Table 2. IGR Performance against Immature Stages of Fungus Gnats

Material	Number of pots	Pretreatment count Number per pot	Post-treatment count (larvae/adults) 7 days	14 days	21 days
Control	15	27.6	84.4	66.3	199.3
Diflubenzuron	15	30.3	7.0	1.0	1.0
Cyromazine	15	27.3	1.5	0.3	0.3
Fenoxycarb	15	29.0	25.5	1.33	3.0

Three flats per treatment, five pots examined per flat, larvae on potato slices recorded after seven and 14 days after treatment, adults on sticky card recorded 21 days after treatment.

Performance Comparisons

Because all of these IGRs exhibited their greatest impact when applied to eggs or early instars of whitefly, timing applications to coincide with this life stage will provide the greatest benefit. These products don't act quickly (with the possible exception of buprofezin against early instars), so assessing control one to three days after application (a common interval with conventional pesticides) will provide misleading information. Instead, you must understand that weeks may elapse before you observe an impact on the whitefly population. By waiting to record adult emergence in the phenoxycarb fungus gnat trial, a true assessment of the material's impact on fungus gnats was obtained. Phenoxycarb affected fungus gnats from pupation to adult emergence phase of development; an assessment based on this IGR's impact on larval survival would have been misleading.

In our studies, only one IGR application was made to a specific life stage, and from this we discovered which whitefly stages will be most affected. In addition, the fungus gnat trial targeted only one stage of development. Because of this, it's dangerous to extrapolate these results to a production greenhouse. Growers are likely to make repeated applications to overlapping whitefly life stages and fungus gnat development stages and generations in a greenhouse. It's possible that two or more applications of these IGRs will contact the same population (this is especially true given the relatively long development time for whitefly) and provide a greater level of control than observed in our trials. In addition, some of these IGRs may impact adults, as has been reported for buprofezin with whiteflies; this wasn't measured in our study. Furthermore, buprofezin has been shown to kill whiteflies via vapor action. This phenomenon was observed in the greenhouse, but wasn't formally investigated.

Multiple applications of these IGRs can be problematic because they aren't immune to resistance development. If used extensively for whitefly or fungus gnat control, resistance will develop. This has been documented in Europe and Israel with whiteflies after the extensive use of buprofezin and pyriproxifen. In a separate study not reported here, we investigated the compatibility of these products with the parasitoid *Encarsia formosa*. As their impact on whiteflies varied widely, these materials differed dramatically in their compatibility with parasitoids.

June 1998.

IPM in Ornamentals: A Guide to Biocontrol

Leslie R. Wardlow

The use of biological controls in integrated pest management (IPM) programs for greenhouse ornamental crops has lagged behind its use in edible crops such as tomatoes and cucumbers. This is because a much wider range of pesticides has been available for ornamentals, where pesticide residues have been less of a problem. Pest resistance to insecticides and acaricides caused the resurgence of biological pest control in edible crops in the early 1970s, and though the same development of resistance was causing problems in ornamentals, there always seemed to be an alternative solution from the wider arsenal of pesticides. In Europe, this all changed when the western flower thrips, *Frankliniella occidentalis*, invaded the European continent from the U.S. in 1983.

This pest's resistance spectrum and its complicated life cycle meant that very intensive insecticide programs were necessary (and still are for unenlightened growers) and that the phytotoxicity and hazards caused by intensive pesticide use became unacceptable to growers.

Growers who have turned to IPM have been better able to deal with the influx of new pests such as the *Liriomyza* leafminers and the tobacco whitefly, *Bemisia tabaci*. No problem for the biological users. Currently, these growers represent about 740 acres of production.

Application Tips

It's really all very simple. About twenty biological controls are commercially available. These are aimed at a range of pests. If you accept that each ornamental crop species has its own pest complex from which it might suffer, then you just make up an IPM program that includes as many biological controls for as many of these pests as possible. Where no biocontrol is available, you have to integrate a pesticide that won't harm your biocontrols. So, in Europe for a crop such as poinsettias, we need only parasitic nematodes against fungus gnats and *Encarsia formosa* against whiteflies; whereas for gerbera, we need to protect against aphids, whiteflies, caterpillars, leafminers, thrips, and tarsonemid mites.

It might sound complicated for gerbera, but it isn't if you accept a basic principle of IPM in ornamentals. Select a basic rate of introduction of the biocontrols that's economically acceptable to "waste." Use this introduction rate on a regular (as little as weekly) basis to keep a reservoir of natural enemies on the crop. These rates are now well tested commercially. The basic rate isn't designed to create a balance of pest and natural enemy, but rather to provide enough natural enemies to attack the likely low numbers of pests that may occur in normal circumstances. Usually, there are enough biocontrols to kill these pests as they arise. In other words, we're putting feeders, not breeders, on the crop.

The technique works well, but immigrations of insects from outdoors into greenhouses are usually too heavy in numbers for the basic rates to cope; in these infrequent cases, extra natural enemies can be introduced or a pesticide may be used. In countries where outdoor climate favors pest reproduction and the risk of immigrations, screening greenhouse openings is essential.

Managing an IPM program

You can use the suggestions in the table below to order the regular delivery of natural enemies. The bugs arrive on a regular basis. Then, you divide them out to the various crops as required. No biological expertise is needed. It's only a matter of putting packets in place throughout the crop, and some crops merely get more packets than others. This is a useful feature of biological control—you don't need the highly qualified staff that's required for spraying pesticides. However, expertise in crop monitoring to check for pests and the presence of natural enemies is required.

Most European growers are now highly competent in this regard. Employees are also trained in pest identification and the symptoms of pest damage so that monitoring can be done as effectively as possible. This has the added effect of improving general crop health, as problems other than pests can be spotted.

You can monitor on a systematic basis, but there are many ways of taking short cuts to make the job quicker and easier. For instance, some crops can be used as indicators of an IPM program's success. Fuchsia will always give early information on whitefly control; cyclamen flowers will tell how thrips control is progressing; hydrangea will warn of the spider mite situation; and ranunculus will always show aphids first. Older crops can predict success or trouble on younger crops, and even weeds can be useful for assessing pest presence. The best indicators are the sticky traps used to monitor flying insect populations.

Many other tips can ensure successful IPM. Growers learn their own thresholds of pests that require action. These vary according to the site and the market for the plants; however, zero-tolerance of pests is possible to achieve, in spite of doubts by some experts. The trick is to anticipate problems. For instance, once any population of aphids develops wings, the battle is lost—a carefully timed aphicide will prevent trouble.

Where Does IPM Work?

IPM is working successfully in pot plants, bedding plants, foliage plants, and nursery stock. There are successes in cut flowers, but generally these growers are reluctant to take up the technique, as open flowers breed thrips so rapidly that the predatory mite *Amblyseius cucumeris* can't keep pace. There are also problems in controlling soil pests, such as slugs, where growers don't sterilize the soil. Nevertheless, the determined flower grower can make IPM work, if only for part of the season. Disease control is sometimes a limiting factor to the expansion of IPM; mildew on roses is a good example, especially if regular sulfur fumigation is the routine control in the nursery. However, there are usually ways around these problems.

Benefits of IPM

Reduced pesticide use is in everyone's interest, especially when working daily in the greenhouse. Growers also like the fact that greenhouses need not be closed up following insecticide use. In addition, you don't have to pay overtime rates for spraying after hours.

Lack of pesticides immediately produces lush, higher-quality plants. Although insecticides may not scorch plants, some cause imperceptible hardening and reduced growth that only becomes noticeable when chemicals are removed.

The IPM grower is in a situation to take advantage of the public demand for a cleaner environment; undoubtedly, more and more legal restrictions are on the way for pesticide use. In Europe, the major supermarkets are already heeding this threat by insisting that their suppliers use IPM techniques, ensuring that their customers have little cause for criticism. Public awareness will increase in this area, and IPM will naturally expand.

IPM was originally designed to deal with the problems of pesticide resistance. It's doing this successfully, and, ironically, it will give a longer life to the pesticides we presently find useful. This is just as well, because the days of cheap, new products are over.

Parasites and Predators for IPM Programs

Target pest	Natural enemy	Rate of use
Thrips and tarsonemid mites	Predatory mites—*Amblyseius* spp. Predatory bugs—*Orius* spp.	Weekly broadcast over crop at five mites per 1. sq. ft. and/or three packets per 10 sq. ft. One per 20 sq. ft. on two occasions at fourteen-day intervals
Aphids	Parasitic wasp—*Aphidius* spp. Lacewings—*Chrysopa* spp. Predatory midge—*Aphidoletes aphidimyza* Fungal disease of insects—*Verticillium lecanii*	One to two per 10 sq. ft. every seven to fourteen days One to two per 1 sq. ft. per week until infestation is cured One to two cocoons per 10 sq. ft. per week Regular spray treatments when humidity exceeds 90%
Spider mites	Predatory mites—*Phytoseiulus persimilis* Predatory midge—*Feltiella acarisuga*	Weekly at one per ten plants 250 cocoons per week until established
Leafminers	Parasitic wasps—*Dacnusa sibirica* and/or *Diglyphus isaea*	250 per week per greenhouse until pest is seen when rate increased to one per ten mines per week
Whiteflies	Parasitic wasps—*Encarsia formosa* and/or *Eretmocerus ca1ifornicus* Disease—*Verticillium lecanii* Predatory beetle—*Delphastus catalinae*	One to five plants per week depending upon risk and whitefly species Regular sprays under high humidities Rates undetermined at present, several introductions recommended for heavy infestations
Caterpillars	Bacterial disease of insects—*Bacillus thuringiensis* Moth egg parasite—*Trichogramma* spp.	Monthly sprays to the whole nursery as a routine Minimum order from supplier at four- to- six week intervals
Mealybugs and scale insects	Predatory beetle—*Cryptolaemus montrouzieri* Mealybug parasitic wasps—*Leptomastix dactylopii*	One to two per 10 sq. ft. per week until established, one per 1 sq. ft. for heavy infestations As above for citrus mealybug
Leaf hoppers	Egg parasite—*Anagrus atomus*	Minimum order from supplier weekly until parasites are established in the crop
Vine weevil and fungus gnats	Parasitic nematodes—*Steinernema* spp. or *Heterorhabditis megidis* Predatory mites—*Hypoaspis* spp.	100,000 nematodes per 10 sq. ft.; these nematodes are also useful against thrips pupae in soil and are known to kill caterpillars and leafminers when sprayed on foliage. 10 per 1 sq. ft.

Screening out Pests

Chris Beytes

One of the most effective ways to prevent insect problems in your greenhouse is to use one of the many types of insect screens now on the market.

The National Greenhouse Manufacturers Association (NGMA) has compiled research data into its *Greenhouse Insect Screen Installation Considerations for Greenhouse Operators.* These standards will help ensure that your screening will keep insects out, let air in and last as long as possible. Following are some screening tips from NGMA's publication.

Selecting Screens

You must consider several factors to match the right type of screen to your operation. There are almost as many types of screening as there are insects, and each has its merits depending on the application. Your challenge is to match the proper screen to your needs. The following points must be considered:

The nature of the insect attack. Does your crop suffer from pests during the entire growing season or just for a limited time? If you need screening only during part of the growing season, you can probably use lighter-duty, less-expensive screening.

Does the crop suffer from a wide range of pests at different times during the season, such as aphids in spring, thrips in summer, and whiteflies in the fall? In this case, you must select a screen tight enough to reduce or exclude as many of these pests as possible.

Type of insects to be screened

Compare insect size to the hole size in the screen to be sure the screen will exclude it. Choose a screen according to the smallest insect you want to exclude. This will also block the larger ones.

Interestingly, it's been proven that thrips (which average 265 microns in size) can be dramatically reduced by screens designed for whiteflies (which range from 565 to 708 microns in size), even though, technically, the thrips are small enough to fit through. This may be because thrips don't recognize screen material as something to feed on, or else they aren't attracted to the white color of most screens.

How will the screen be used?

Light duty: You can use a screen inside the greenhouse to create a pest-free zone. Or you may only need to have the house screened for certain times or seasons. For example, thrips may only be a problem in late spring, allowing you to take down the screen in the summer, fall, and winter.

Heavy duty: Screens that will be up year-round will be exposed to sun, wind, hail, rain, and snow. This requires stronger, more rigid screens. If installed as roll-up screens, they need to endure rolling and unrolling. If used near a high-traffic areas, they need to endure wear and tear from workers and equipment brushing by. If you need weights in the screen to keep it from interfering with vent operation, the weights can cause wear.

Screen Types

Different types of screen material are available, including stainless steel and brass (durable but expensive), polyethylene and polyethylene/acrylic (strong, long lasting), and nylon (less expensive, less airflow).

Whether applied inside (left) or outside (right), the proper insect screen stops pests while allowing good airflow.

Woven screen is the most common construction method and is a good trade-off between hole size and airflow. Knitted screens resist tearing and unraveling but may cause greater air restrictions. Film screening, which is a sheet of film with punched holes, is most restrictive to airflow. It must also be applied with the correct side facing out.

Maintenance

It's best to size your screen for proper airflow with the help of your screen supplier and your greenhouse builder. They can accurately check static pressure, air exchange rates, and other factors.

Once you have a good screen installation, you have to keep the screen clean to prevent airflow problems. Do that by using water from a standard hose and nozzle, spraying from inside the screen toward the outside. Don't use a high-pressure sprayer or brush—you can alter the hole size.

Also, clean with the ventilation fans off—water can block the screen holes, completely stopping airflow.

NGMA standards are available from the NGMA office, Tel: (800) 792-6462; outside of the U.S., Tel: (303) 798-1338.

Pest Control, March 1999.

Cinnamite: The Ideal Reduced-Risk Pesticide?

Michael P. Parrella, Dave von Damm-Kattari, Gina K. von Damm-Kattari, and Brook C. Murphy

Imagine a pesticide with the following attributes: no mammalian toxicity, a four-hour reentry interval, a zero-day preharvest interval, and the ability to kill many common greenhouse pests including adult and immature western flower thrips, aphids, whiteflies, and the egg and motile stages of spider mites. Add to this the capacity to repel certain insects and the capability of controlling key fungal diseases on floriculture crops. To top it all off, there is a pleasing aroma associated with its application. Now, suppose we said a product exists that has these attributes, is registered under the trade name Cinnamite, and obtained a federal registration in February 1999. The material is currently available nationwide as a 30% flowable formulation of cinnamic aldehyde and is marketed by Mycotech Corporation in Montana.

As part of an ongoing project, our research program is evaluating numerous reduced-risk pesticides in an attempt to find effective products. Ideally, these products may be more conducive to the tenets of integrated pest management and may be able to replace traditional organophosphate and carbamate pesticides slated for removal according to the Food Quality Protection Act. Cinnamite is one of these reduced-risk products.

Reduced-Risk Pesticides

The term *reduced-risk* is applied to a broad range of pesticides, but the label must be officially given to a pesticide by the Environmental Protection Agency (EPA). Typically, reduced-risk materials are those that, compared to existing alternatives, result in reduced risks to human health and the environment. The EPA issued Pesticide Registration Notice PR 97-3 to pesticide manufacturers and registrants in September 1997 with new guidelines for applying for expedited review and approval of reduced-risk pesticides.

The major advantage to the registrant of a reduced-risk pesticide is the priority EPA gives these products for review. The average time EPA needed to review and register a conventional pesticide in the 1994-95 time frame was about thirty-eight months; for reduced-risk pesticides the average time was about fourteen months, providing a distinct advantage in the marketplace. The owners of Cinnamite (ProGuard Inc., Suisun City, California) took advantage of the reduced-risk status of their product and received a federal label (and subsequent state labels) in record time. Mycotech Corporation, specializing in natural pest control solutions, has since been licensed to distribute Cinnamite.

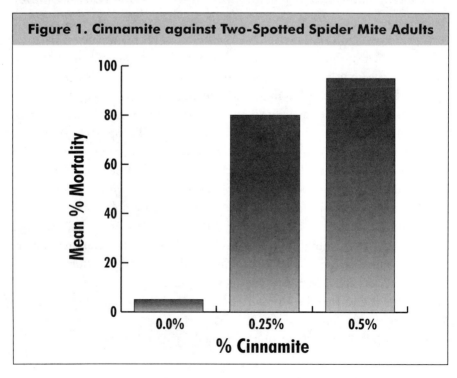

Figure 1. Cinnamite against Two-Spotted Spider Mite Adults

It isn't unusual for reduced-risk pesticides such as Cinnamite to have some of the following characteristics: short persistence, brief reentry interval, low toxicity to mammals and fish, safety to nontarget organisms, target specificity, compatibility with natural enemies, activity at low dosages, and lack of potential to contaminate soil and ground water. Rarely will one pesticide exhibit all these, but a remarkable number of these favorable characteristics are found in Cinnamite.

Reduced-risk products can come from many different sources, including natural products, synthesized analogs of naturally occurring biochemicals and microbial control agents, among others. Cinnamite is the synthesized analog of naturally occurring cinnamon oil, which has been known for some time to have insecticidal/fungicidal properties. However, cinnamon oil and its synthesized analogs (which are better known as food additives), haven't been exploited for pest control until now. We've been evaluating Cinnamite in extensive studies during the past three years and first introduced this product to the floriculture industry during the grower tour that was part of

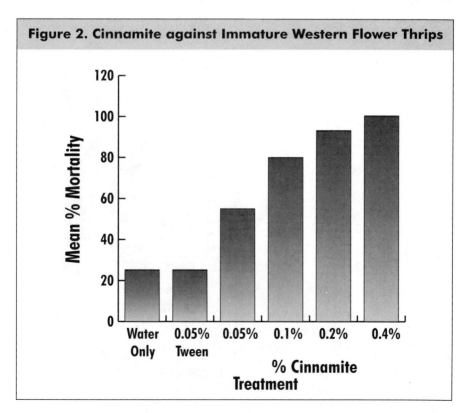

Figure 2. Cinnamite against Immature Western Flower Thrips

the Conference on Insect and Disease Management on Ornamentals (sponsored by the Society of American Florists) and held in San Francisco, California, in February 1996. Cinnamite has come a long way since then.

Putting It to the Test

Our testing of this product began in fall 1995, when ProGuard Inc. approached us with this potential new insecticide. They presented us with evidence that demonstrated it was insecticidal, exhibited some repellency, and was active against a range of plant pathogens. We started evaluating Cinnamite with a range of dosages against several common greenhouse pests. Initial studies were very encouraging; the material demonstrated good activity against whiteflies, western flower thrips, melon aphids, and the motile stages and eggs of spider mites.

After this broad initial screening, we zeroed in on a few pests and developed dose-response lines for each; this provided us with a more detailed picture of how pest mortality increases with increasing dosage. Ultimately, studies such as these can help us better pinpoint the field-use rate that will

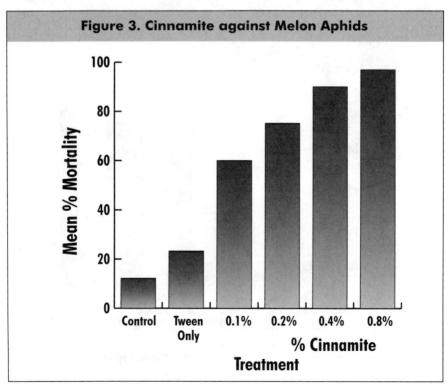

Figure 3. Cinnamite against Melon Aphids

be on the label. Acceptability of the product hinges on its performance and compatibility with the target crop—phytotoxicity isn't acceptable when dealing with aesthetic-value ornamental plants. We did notice a few problems on poinsettia, for example, but this appeared to be a formulation problem rather than a problem directly caused by the active ingredient.

Based on laboratory and greenhouse phytotoxicity trials, we concluded that a concentration of 0.2 to 0.25% would be an acceptable balance between efficacy and phytotoxicity. At this concentration, Cinnamite's performance is as follows: approximately 80% control of adult mites (see fig. 1), approximately 100% control of mite eggs, approximately 75% control of immature western flower thrips (see fig. 2), less than 50% control of adult thrips, and approximately 80% control of melon aphids (see fig. 3).

Armed with this performance data, we went into commercial greenhouses and set up large phytotoxicity trials on cut flowers and potted flowering plants. The details of one of these trials are outlined below.

The grower provided us with eighteen common nursery crops grown at the facility. A total of 270 plants were divided into approximately five plants per cultivar per treatment, resulting in roughly ninety plants per treatment. Phytotoxicity was evaluated forty-eight hours after the spray. (Refer to table 1 for the list of methods of quantifying phytotoxicity for each crop.)

We briefly noted the effect on open flowers, flower buds, and leaves. Phytotoxicity was evaluated by counting the total number of leaves on each plant and recording the number showing signs of phytotoxicity. Pictures were taken of all the plants.

The spray was repeated one week after the initial spray, and the effects were again recorded forty-eight hours afterward. Due to high levels of phytotoxicity, the 0.5% treatment wasn't sprayed a second time. Data on phytotoxicity were similar after one or two applications, so results are summarized for the first treatment (table 1).

Examining figure 1 reveals no phytotoxicity at 0.25% on *Euphorbia milii,* heather, ivy, potted chrysanthemum, lavender, *Manettia,* rosemary, *Streptocarpus,* and thyme, and minor phytotoxicity on *Bouvardia,* cyclamen, gerbera, ornamental pepper, passion flower, oriental lily, and small chrysanthemum. Unacceptable injury was recorded on mini roses.

The results from this phytotoxicity trial were less than what we would hope for in a product with enormous potential to impact a broad range of insect and mite pests attacking ornamental crops. However, it's clear that this

Table 1. Quantifying Phytotoxicity

Crop/cultivar(s):
Mini Roses: Frosty, Sunset, Orange
Ornamental Peppers: Masquerade, Holiday Flame
Lavender
Kalanchoe: Tonorio, Revelry, Kerinici
Bouvardia: Stephanie
Passion Vine: Lavender Lady, Incense
Thyme
Cyclamen: Laser series
Chrysanthemums: Shasta, Summertime and two other varieties
Oriental Lilies: Stargazer
Jasmine
Ivy: California
Gerbera: Festival, Floripot Maxi Mix
Heather
Manettia inflata: Bicolor
Streptocarpus: Hera, Ulysses, Thalia, Minerva, Demete
Euphorbia millii: Gondula, Gabriela
Rosemary

Treatments:
Water only control, 0.25% Cinnamite with 0.1% Tween, 0.5% Cinnamite with 0.1% Tween

Dates and times of applications:
July 22, 8 a.m.
July 30, 8 a.m.

Application method:
30-gal. cart direct-driven commercial sprayer

Volume sprayed:
To drip, 2.5 to 4.0 gal. of material sprayed per treatment plots

Irrigation method/soil moisture:
Direct soil application/saturated soil

Experimental schedule:
July 14	Plants selected and arranged in plots.
July 15	Treatments applied.
July 24	Observation and photography of plants.
July 30	Second application of control and 0.25% treatments.
August 1	Final observation and photography of plants.

product has the potential to become an important tool in the pest control arsenal for many floriculture producers. ProGuard Inc. has developed a new formulation of the material promising fewer phytotoxicity problems. This formulation is currently being marketed in forty-seven states and should be available in all fifty states by summer of 1999.

June 1999.

Bugs Not Interested in Sunblock?

Brandi D. Thomas

Using UV-absorbing polyethylene film in your greenhouse may prove to deter destructive insects from entering, according to a report by the California Ornamental Research Federation (CORF).

Studies conducted by Drs. Heather Costa and Karen L. Robb (of the Dept. of Entomology at UCR and the University of California Cooperative Extension, respectively) involved the use of two small experimental tunnels—one made of regular plastic, the other of high UV-absorbing plastic.

In the experiment, both silverleaf whiteflies and western flower thrips were released into a box at the center of the two tunnels, having the choice to enter either one or the other. Both types of insects seemed to exhibit a preference for the standard, non-UV tunnels, as they retained 85 to 94% of the silverleaf whiteflies and 90 to 98% of the western flower thrips.

These findings formed the basis for a further ongoing study in which naturally occurring populations of insects and pathogens have been released into three hoop houses, constructed of standard plastic, a high UV-absorbing plastic with insulator component and a standard plastic with insulator component. Each hoop house contained two beds of chrysanthemum and two of solidago.

The numbers of insects found in each greenhouse are being recorded by three-by-five-inch yellow sticky cards and through plant samplings. So far, it appears that the numbers of thrips and aphids being caught are lower in the UV-absorbing greenhouses, while there has been no difference in the numbers of trapped whiteflies. Expanded studies will include planting crops of *Lisianthius* for more conclusive results.

Culture Notes, August 1999.

Banker Plants Keep Predators Productive

Techni-Gro Greenhouses in Chilliwack, British Columbia, Canada, is working hard to grow 60,000 square feet of cut gerbera production using 100% biological controls.

For aphid control, owner William Stolze uses three predators: *Aphidius colemani, A. ervi,* and *Aphidoletes aphidimyza.* But what do these hungry predators eat when they've gone through all the aphids in the gerbera crop? They eat aphids living on the six banker plants hanging in each 30,000-square-foot section.

Banker plants serve as hosts for an insect that can serve as a meal for predators when they can't find anything in the production crop. In this case, the banker plant is wheat, and the aphid is a type that only feeds on monocots—that keeps them from infesting the gerbera, which are dicots.

At first, Techni-Gro released the predators right into the banker plants but found they wouldn't disperse well. Now they put them in the gerbera crop and let them make their own way to the banker plants.

Christine Koch, extension specialist with the British Columbia Ministry of Agriculture, says that four or five other gerbera and rose growers are trying biological controls, as are some growers who specialize in propagation.

Biocontrol isn't easy, but with help and regular scouting from Koppert Biological Systems, which supplies their biological controls, Techni-Gro has reduced sprays to the occasional application of Dipel for worms, while a dusting of sulfur controls powdery mildew. William hopes to soon promote his 100% biological status on every shipment.

Culture Notes, December 1999.

Pest Connection

Raymond A. Cloyd and Karen K. Rane

Insect and disease problems are very much interrelated, and managing one may have an impact on the presence or severity of another. Sometimes the interaction is very specific and is required for disease to occur, while in other instances the relationship is a casual association. In either case, managing insects in your greenhouse may not only reduce injury directly due to these pests but may also minimize the potential for disease.

Here's a look at five major greenhouse pests and the diseases they can transmit.

Aphids

While aphid-borne virus diseases aren't extremely common in greenhouse ornamental production systems, some important viruses, such as cucumber mosaic virus, dasheen mosaic virus, and tobacco ringspot virus, are occasionally found causing symptoms in crops such as New Guinea impatiens, geraniums, begonias, and various foliage plants.

Aphids acquire virus particles as they feed on infected crop plants or weeds. Virus particles are then transmitted to new plants through the aphid's piercing-sucking mouthparts. The relationship is

More than just a nuisance on their own, aphids are also the most common vectors of many viral diseases in the field and in greenhouses. *Photo by Raymond A. Cloyd.*

usually specific—that is, a particular virus can be transmitted by only one or a few species of aphids. Most aphid-transmitted viruses that infect ornamental plants need only a relatively short feeding period (anywhere from ten seconds to several minutes) for the insects to obtain the virus, and transmission occurs relatively quickly as well. Viruses that occur in field-grown plants may cause problems in greenhouse crops if field aphids enter the greenhouse through open doors, vents, or sidewalls. Aphid-borne viruses can become perpetuated in greenhouses if weeds are allowed to grow and serve as sources of virus particles.

Whiteflies

Spores of potential fungal pathogens or bacterial cells can stick to the external body surfaces of whiteflies and be carried with them as they fly. The pathogens can then be deposited on plant surfaces as whiteflies are feeding, and the pathogens invade plant cells that are damaged by such feeding activities.

In recent years, whiteflies specifically have been shown to vector viruses known as geminiviruses. At the present time, these viruses are known to cause economic damage primarily in vegetable crops such as tomatoes. However, a few of these viruses are known to infect ornamentals, so the potential exists for future disease problems in greenhouse ornamental production systems.

Thrips

Thrips are capable of vectoring two destructive viral pathogens: tomato spotted wilt virus (TSWV) and impatiens necrotic spot virus (INSV). Both are classified as tospoviruses. Eight species of thrips (including western flower thrips) are capable of transmitting TSWV. However, only western flower thrips, *Frankliniella occidentalis*, is a known vector of INSV.

Thrips acquire viruses as larvae by feeding on virus-infected plants or weeds. Weeds that are known to harbor these viruses include oxalis, chickweed, hairy bittercress, prostrate spurge, and jewelweed. The larval stages of thrips acquire a virus by feeding on an infected host; virus acquisition takes approximately fifteen to thirty minutes. The virus is then transmitted when adult thrips feed, generally requiring as little as five minutes.

Viruses may multiply inside a thrips' body, but there doesn't appear to be any transovarial transmission (the virus isn't passed on to offspring or young). First-instar larvae (zero to two days old) are better able to acquire a virus and are more efficient vectors as adults compared with second-instar larvae. The ability to transmit a virus declines when it's obtained at a later stage, and adult thrips are unable to acquire the virus during feeding.

Fungus Gnats

Fungus gnats, *Bradysia* spp., damage plants directly through larval feeding or indirectly by transmitting diseases. Both the adult and larval stages are capable of disseminating and transmitting diseases to ornamental crops. Adults can carry on their body viable spores of *Botrytis cinerea*, *Verticillium alboatrum*, *Thielaviopsis basicola*, and *Fusarium oxysporum*. Adult fungus gnats may acquire spores when they emerge from the growing medium and/or when they move around on the medium surface. Although fungus gnat adults aren't strong fliers, they can readily move from pot to pot within a crop. If spores fall off an adult's body and the environmental conditions are conducive for spore germination, disease may develop.

Fungus gnat larvae can transmit diseases directly and indirectly. Larval feeding on roots can create wounds that allow secondary, soilborne pathogens such as *Pythium* or soft-rot bacteria to enter. In addition, larvae may deliver spores of *Pythium* spp., *Thielaviopsis basicola*, and *Cylindrocladium* spp. directly to feeding wounds. Larvae often carry *Pythium* and *Thielaviopsis* spores inside their bodies; oospores (the sexual stage of water-mold fungi) can remain in the insects' gut throughout their entire life.

Larvae are able to ingest and excrete viable spores; this, combined with their direct feeding, increases a plant's susceptibility to disease-causing organisms.

Shore Flies

Research has shown that shore fly adults and larvae are capable of transmitting *Thielaviopsis basicola* in their frass (excrement). In addition, shore fly adults can disseminate the water-mold fungus *Pythium aphanidermatum* and the bacterial pathogens *Erwinia carotovora* and *Pseudomonas cichorii*. The larvae acquire plant pathogens in the root zone of plants. These pathogens are retained through pupation and into adulthood. Adult shore flies can also obtain fungi directly by ingesting the spores. Adult shore fly frass deposited on the lower leaves and stems of susceptible plants may lead to host infection if proper environmental conditions are present.

Crop Management

Greenhouse managers who understand the relationships between insects and diseases can focus their management strategies to deal with these insect pests early before they cause a bigger problem by transmitting diseases. Similar to many pests, this requires the implementation of a holistic pest management program, which uses scouting and physical, cultural, chemical, and biological management strategies to produce a quality crop.

August 2000.

Garden Mum Insect Control

Cassy Bright and Roberto Gonzalez

Garden mums are a grower favorite, but they're also popular with a number of insects, which can seriously damage the crop. Fortunately, there are many options to help you minimize damage and maximize profits.

Insects most likely to be found on garden mums include aphids, leafminers, thrips, whiteflies, mites, and fungus gnats. While all of these can be serious and damaging, aphids, leafminers, and thrips account for the majority of insect problems you'll encounter.

Aphids

Of these three, aphids tend to occur most frequently and in the greatest quantity, although they're rarely responsible for major aesthetic damage.

There are several kinds of aphids that can attack a garden mum crop. The most common is the green peach aphid, *Myzus persicae*. It's important to correctly identify this pest before beginning a spray program because this aphid has the greatest resistance to pesticides.

The next most common aphid on garden mums is the melon (or cotton) aphid, *Aphis gossypii*. Usually dark in color but ranging from light green to nearly black, melon aphids tend to attack later in the crop, typically around the bud stage. Early detection and control is important as they can cause significant damage to flowers.

Aphid control options include traditional chemical sprays, insecticidal soaps and oils, physical barriers, and predatory insects. Traditional chemical sprays can be used effectively when rotated carefully. Orthene, Mavrik, and Avid sprays can be used in a rotation, as can insecticidal soaps and oils. Frequently used insecticidal oils and soaps include Safer Soap, Saf-T-Side, Sunspray Ultrafine, and M-Pede. Don't use these once buds are showing color because they may damage flowers. For heavy aphid infestation, Thiodan can be used once or twice.

Predatory insects such as ladybugs, *Hippodamia convergens*, and their larvae can be used to reduce populations. Often growers incorporating biological control will tolerate a low level of infestation in the first stages of the crop and use a chemical spray for eradication before shipping.

For added aphid prevention, physical barriers such as screens over greenhouse vents or as wind barriers outside the greenhouse can be useful. Border plantings of natural aphid repellents such as marigolds are another option.

Leafminers

Leafminers can have a devastating effect on garden mums. These insects burrow tunnels in leaves, permanently scarring leaf tissue. Avid, Citation, and many pyrethroid insecticides work effectively against them.

A physical ground barrier is an effective method to prevent further infestations. Leafminer larvae emerge from the leaf and drop into the soil to pupate and develop into adult flies. Blocking soil with a groundcover such as saran, plastic, or hydrated lime will prevent the larvae from continuing its life cycle. This can significantly reduce the numbers of adult flies laying eggs inside leaves. Yellow sticky traps are a good method to monitor fluctuations in adult leafminers populations.

Thrips

Thrips cause major damage on garden mums by deforming flower buds and petals. Monitoring thrips populations early in the crop cycle is essential to avoid later losses. Sticky traps are used to monitor populations; however, because of the reclusive nature of these insects, populations can often be much higher than the cards indicate. Effective monitoring includes careful examination of plants using a hand lens.

Dursban and Grand Slam are recommended for thrips control. Early detection and action are essential—once thrips are thriving inside the buds, it's very difficult to control the infestation, as chemical penetration is poor.

Less Common Pests

Some of the less common garden mum pests are whiteflies, mites, and fungus gnats. For whiteflies, begin with prevention by keeping the growing area clear of weeds. If an infestation does occur, Pounce, Malathion, Mavrik and insecticidal soap can be effective treatments.

Mite problems occur during hot, dry weather and when plants are under stress. Heaviest infestations can lead to webbing covering the plant canopy. Avid, Mavrik, and Pentac (use flowable form for outside use) are effective miticides. If feasible, occasionally wetting the plants' foliage can reduce mite populations.

Fungus gnats can be a serious problem in propagation if left unattended. The adult flies seldom cause damage to garden mums, but the larvae can severely damage roots. Oxamyl granules offer excellent control of larvae while Vydate sprays are effective on the adult flies. Fungus gnats can also be monitored with yellow sticky traps.

Pest Control, August 1997.

On-Target Insect Control for Herbs

Whether you're producing basil, oregano, mint, or tarragon, the success of your herb production depends on a targeted pest-control program. Insects and mites can cause some of the worst damage to herb crops. Knowing which chemicals work best to control which pests is half the pest-control battle. Here, courtesy of Tina Smith, University of Massachusetts Cooperative Extension specialist, is our guide to controlling insects and mites on herbs.

Common name	Trade Name	Use	Pests controlled
Azadirachtin	Azatin	Greenhouse, outdoor	Aphids, beetles, caterpillars, fungus gnats, shore flies
Bacillus thuringiensis Subsp. CryIA, CryIC	Mattch	Greenhouse, outdoor	Caterpillars
Bacillus thuringiensis Subsp. *kurstaki*	MVP II, Dipel 2X	Greenhouse, outdoor	Caterpillars
Bacillus thuringiensis Subsp. *israelensis*	Gnatrol	Greenhouse, outdoor	Fungus gnats
Beauveria bassiana	BotaniGard	Greenhouse, outdoor	Aphids, mealybugs, thrips, whiteflies
Insecticidal soap	M-Pede	Greenhouse, outdoor	Aphids, leafminers, leafhoppers, mites, thrips, whiteflies
	Insecticidal soap	Outdoor	Aphids, leafminers, leafhoppers, mites, thrips, whiteflies
Horticultural oil	Sunspray Ultra-Fine oil spray	Greenhouse, outdoor	Aphids, beetles, leafminers, leafhoppers, mites, thrips, whiteflies
Pyrethrins	1100 Pyrethrum TR	Greenhouse	Aphids, beetles, caterpillars, fungus gnats, leafhoppers, mealybugs, mites, scales, thrips, whiteflies

Culture Notes, April 1999.

Insect Pests of Perennials

Joanne Lutz

With the wide array of perennials grown today, it's critical to come up with control strategies before extensive damage occurs. Crucial keys for pest managers include early detection, knowledge of key pests, and distinguishing symptoms from the damage produced by an insect's mouthpart. Time of year, location on a plant, and thresholds are other key factors influencing control options. Insects can be classified by mouthpart (either chewing or piercing-sucking) and can either feed internally between the leaf tissue or bore into stems and crowns.

Chewing Insects

Chewing insects such as caterpillars and sawflies can cause extensive damage when feeding on leaf surfaces. Look for frass (debris or excrement produced by insects); turn leaves over and inspect carefully. Perennials that are frequently damaged by caterpillars and should be monitored closely include *Artemisia* 'Silver Brocade', polygonum, *Heuchera* 'Autumn Bride', monarda, hemerocallis, pulmonaria, primula, and *Asclepias*. Also, carefully monitor hibiscus and ferns for sawflies. Horticultural oils, insecticidal soap, and *Bacillus thuringiensis* (DiPEl) can be used on eggs and early-instar caterpillars. *Beauveria bassiana* (Naturalis-O), neem pesticides (azadirachtin), or insect growth regulators such as dimilin (Adept) are good biorational pesticides to use in the greenhouse on larvae. Pyrethrums have quick knockdowns and short residuals, while synthetic pyrethroids such as Talstar or Decathalon have longer residuals, work on contact, and are broad-spectrum pesticides. Adults are only controlled if they ingest or come in contact with the pesticide residue.

Japanese or Oriental beetles, black vine weevils, and grasshoppers cause damage such as leaf chewing, leaf notching, or skeletonized leaves and flowers. Beetles are attracted to aster, astilbe, gaillardia, hibiscus, *Oenothera*, rudbeckia, polygonatum (silver lace vine), and phlox. Black vine weevils at nursery operations can cause significant damage and will become a serious pest if left unchecked. Adults are difficult to find, but the damage they leave behind isn't. They feed at night on outer leaf margins and leave a notching pattern that can reduce perennials' aesthetic quality. However, it's the larvae that can cause significant damage—not by feeding on leaves, but on roots. Astible, *Bergenia*, sedum, *Tricyrtis*, polygonum, *Rodgersia*, *Filipendula*, *Helleborus*, hosta, and *Liriope* seem to be some of the weevils' favorites. The beneficial nematode *Heterorhabditis bacteriophora* (such as the cruiser nematode) is an excellent biological pesticide and should be applied when soil temperatures are between 60° to 90°F. Marathon (imidacloprid) granules, Orthene, or Turcam at label intervals can reduce adult feeding.

Slugs and snails can cause irregular chewed holes in leaves and deposit unsightly slime trails. These night feeders can defoliate and destroy tender young seedlings. During the day, they can be found under pots and flats. Slugs favor *Belamcanda*, tradescantia, aster, and *Iris cristata*, while snails prefer *Achillea*, campanula, viola, delphinium, hemerocallis, hosta, gypsophila, dianthus, *Papaver*, and *Saponaria*. Control is best achieved by reducing conditions that attract moisture: Provide wider spacing to allow for good air circulation, and monitor watering practices that allow foliage to be dry by evening. Products such as diatomaceous earth or slug baits are effective if applied at label intervals.

Leafhoppers can cause twisted foliage, chlorotic leaves, and discolored or deformed flowers. Look on leaf undersides for nymphs or adults and their shed skins. Perennials that leafhoppers find attractive include aster, *Oenothera, Tanacetum,* centaurea, coreopsis, phlox, rudbeckia, wisteria, gypsophila, *Gaura,* and *Helianthemum.* Repeated applications of insecticidal soap and insect growth regulators or systemic pesticides are necessary because of the need to control multiple generations.

Boring Insects

Boring insects cause damage when larvae enter stems or other plant parts and feed inside, causing decline, dieback, or death.

The European corn borer and stalk borer are the two most common, although many species of beetles are also known as borers. Monitor aster, chrysanthemum, delphinium, gaillardia, dianthus, phlox, rudbeckia, *Persicaria,* lychnis, and salvia. Removing infected stems and sanitizing in fall will help reduce overwintering. Chemical pesticides generally aren't effective once the borers are inside the stems.

Leafminers are mining insects whose larvae feed within the leaves of plants, causing light green or white serpentine trails that turn brown. Larvae eventually emerge from the leaf; however, the damage is already done. Susceptible perennial plants include aquilegia, senecio, chrysanthemum, salvia, veronica, delphinium, gypsophila, and *Aconitum.*

Piercing-Sucking Insects

The piercing-sucking group includes thrips, aphids, whiteflies, and mites. They have sharp, needlelike mouthparts that pierce plant cells and draw the sap, leaving signs of small specks, chlorotic spots or streaking, often with some twisted, stunted, or curled growth. Often, a toxic reaction occurs from the saliva of some insects, which can be damaging to the perennial.

Thrips can be the most economically destructive pest because of the known transmission of the plant-damaging tospoviruses INSV and TSWV from at least three thrips species. The western flower thrips is the most prevalent species found throughout the U.S. The larvae feed on an infected plant, then transmit the virus to other plants through feeding. Look for tiny black fecal spots and silvery streaking of foliage (caused by the rasping-sucking mouthpart) when monitoring. Perennials to scout for thrips damage include *Papaver, Lupinus,* hemerocallis, *Platycodon,* phlox, verbena, geranium, *Asclepias,* tradescantia, and *Alcea.* Herbaceous perennials found with the tospovirus include lobelia, *Stokesia,* hosta, *Tricyrtis, Helleborus, Physostegia,* and polemonium. Biological controls of thrips include predatory phytoseiid mites, minute pirate bug (*Orius* spp.) and the entomopathogenic fungus *Beauveria bassiana*. Chemical pesticide choices for control include Conserve, Azatin XL, Decathalon, Orthene, DuraGuard, or Talstar. Multiple applications five days apart are often necessary for control.

Aphids are frustrating pests due to their high reproduction rate and resistance to most pesticides. Green peach aphid, melon aphid, chrysanthemum aphid, and foxglove aphid are the most common, although many other species are found on perennials and grasses. Good indicators of damage include twisted, stunted new growth and the presence of shed skin and honeydew, as aphids also serve as vectors of viruses. Aphid perennial favorites include *Achillea*, aquilegia, chrysanthemum, digitalis, dianthus, viola, sedum, *Oenothera*, monarda, *Alcea*, ajuga, salvia, and *Hypericum*. Effective biological control choices include ladybird beetles, green lacewings, aphid midges, parasitic wasps, and *Beauveria bassiana*. Biorational pesticides include insecticidal soaps, horticultural oils, and insect growth regulators.

Common mite pests include the two-spotted spider mite, cyclamen mite, and broad mite. Fine webbing, stippled foliage and yellow or bronze leaves are indicators of activity of the two-spotted spider mite. Perennials mites find attractive include buddleia, scabiosa, *Knautia*, salvia, lobelia, *Caryopteris*, verbena, geum, *Alcea*, viola, *Asclepias*, and *Crocosmia*. Miticides for two-spotted spider mites include Sanmite, Hexygon, Avid, Floramite, and Cinnamite. Horticultural oils can be used as a control. Although they're invisible to the naked eye, cyclamen and broad mites' damage is identified by twisted, curled, brittle, darkened, or scabby new growth. Stachys, delphinium, *Aconitum*, verbena, clematis, viola, and chrysanthemum are favorites of this mite. Cyclamen and broad mites require repeat applications of Kelthane, Talstar, or Thiodan.

Whiteflies are generally greenhouse pests but can become nuisances outdoors on host plants. The most common species include greenhouse whitefly, silverleaf whitefly, and the bandedwinged whitefly. Yellowing, reduced vigor, or tiny, white flying insects when plants are disturbed are key symptoms of whitefly. Honeydew and sooty mold are also associated with whitefly populations. Perennials whiteflies find attractive are hibiscus, *Eupatorium*, veronica, salvia, *Nepeta*, rudbeckia, *Alcea*, *Boltonia*, aster, and verbena. Biological control includes *Beauveria bassiana* and *Encarsia formosa*. Horticultural oils, insecticidal soaps, or insect growth regulators work well when thoroughly applied to lower leaf surfaces. Many chemical pesticides are labeled for whitefly control, including Tame, Orthene, Topcide, and Resmethrin.

Good sanitation, inspection, and quarantine of incoming plants will help reduce the introduction of unwanted insects into your greenhouse and nursery. Before applying controls, know the key pests and the appropriate pesticides to apply at the most vulnerable stage of the life cycle. Avoid stressful conditions, and follow a nutrient management program to provide optimum growing conditions to give your plants a better chance to tolerate insect attack.

Pest Control, June 1999.

Chapter 6
Specific Pests
Aphids

Aphids in the Greenhouse
John P. Sanderson

Aphids are common pests on many greenhouse crops. Their mere presence can decrease the value of a plant. These insects feed by inserting their stylet-like mouthparts through plant tissue directly into the phloem and removing plant sap. They're commonly found on leaf undersides or on stems or buds. Some species infest roots. Their feeding can reduce plant vigor and lead to stunting and deformities. Their sticky honeydew, which results in sooty mold, and the skins they leave behind after they molt can make plants look ugly. Aphids found on open blooms will ruin sales. Aphids transmit about 60% of all plant viruses on agricultural crops worldwide.

Identification
At least twenty-five species of aphids differing in size, coloration, and food preferences can infest greenhouse plants. They're small (1 to 4 mm), slow-moving, soft-bodied, and pear-shaped without obvious segmentation into head, thorax, and abdomen. Their legs and antennae are typically long and slender. Most notably, aphids have a pair of unique structures called cornicles that resemble "tailpipes" near the end of their abdomen. Adults may or may not have wings.

Knowing the species of aphid present in an infestation can be important in achieving control. Some pesticides, and especially some natural enemies, work better on some species than on others.

The most common species by far are the green peach aphid, *Myzus persicae*, and the melon or cotton aphid, *Aphis gossypii*. Both species can infest a wide variety of floral crops. The color of green peach aphids can vary from light green to rose. Melon aphids are smaller and may vary from yellow to light green to very dark green. The cornicles of green peach aphids are light colored with a black tip, while melon aphid cornicles are uniformly very dark. Your cooperative extension service or a diagnostic laboratory can assist you in identifying most aphid species.

Biology

Under most greenhouse conditions, all aphids of an infestation are females, and each female gives live birth to more females without the need to mate or lay eggs. Their ability to reproduce without mating or egg production can cause their populations to increase explosively, especially because individuals can mature and begin to reproduce in as few as seven days under ideal temperatures.

Aphids will often actively search for soft, young plant tissue that is nitrogen-rich. Green peach aphids are most common on the upper leaves and stems of a plant. Melon aphids are usually found throughout a plant, making thorough pesticide coverage important for control.

Monitoring

Aphid control is much more successful when aphids are detected and controlled early. As with all pests, you should use a routine scouting program. Yellow sticky cards may provide an early indication of an aphid flight into the greenhouse from outdoor sources, particularly in the spring or fall. However, it's even more important to inspect plant foliage for the more common unwinged forms on at least a weekly basis.

Check several plants on each bench throughout the greenhouse. Aphids can be spread on clothing, so check plants along walkways or near doorways. Look at buds, stems, and lower surfaces of leaves. Look for white cast skins, honeydew, deformed leaves, and ants that may be searching for honeydew. A greenhouse map of infested areas will help target areas to be sprayed and monitored.

Management

Carefully inspect plant material brought into your growing area; don't purchase infested plants or cuttings. Aphids often prefer certain plant cultivars to others. Ask your supplier for information about which cultivars are more susceptible to aphids, or keep your own records to aid in cultivar choice. Keep weeds eliminated within and near your greenhouses—these weeds can serve as a reservoir for migrating or ant-carried aphids.

Insecticides that have systemic or translaminar properties tend to be more effective for aphid control than contact insecticides, provided that a sufficient amount of insecticide reaches the aphid feeding sites. However, contact insecticides can be very effective. Thorough spray coverage and canopy penetration is important. One to three sprays of a contact insecticide at five- to seven-day intervals with good coverage should provide control.

Differences in insecticide resistance among aphid populations make it difficult to generalize about the effectiveness of a particular insecticide, but you can evaluate the effectiveness of various pesticides under your own conditions by recording scouting information. To aid in evaluating insecticide effectiveness, mark several infested plants with flags or flagging tape and record an estimated number of living aphids on each. Several days after you apply an insecticide, record the number of surviving aphids again. Examine plants carefully and frequently to determine if repeat applications will be needed.

Many natural enemies are available for aphids, including green lacewings, parasitic wasps (*Aphidius colemani*), predaceous midges (*Aphidoletes aphidomyza*), and insect-pathogenic fungi (*Beauveria bassiana*, the active ingredient in BotaniGard and Naturalis-O, and *Paecilomyces fumosoroseus* and *Verticillium lecanii*, which may eventually be available in the U.S.). Aphid biological control will be more successful if you use combinations of appropriate natural enemies or compatible insecticides with the natural enemies. For success, it's essential that you monitor and integrate biological control with cultural or physical control tactics.

Pest Control, April 1998.

Fungus Gnats

Stopping Disease-Carrying Fungus Gnats

Jim Willmott

Fungus gnats are the greenhouse pests that most deserve the Rodney Dangerfield Award: They just don't get enough respect! Too many growers aren't aware of the serious damage they can cause. In recent years, fungus gnats have become more than just a nuisance.

Besides direct feeding damage by the immature or larval stages, fungus gnats carry and spread pathogenic fungi, including *Pythium* and *Thielaviopsis*—two of the most serious root-rotting pathogens afflicting greenhouse crops in recent years. Other fungal pathogens may be spread as well.

Fungus gnats also present a serious quality issue. It's simply unacceptable to ship fungus gnats to the consumer—especially with crops that are destined for indoor display in a home or commercial situation. Poinsettias, bulbs, and other pot and foliage plants too often reach their destination with fungus gnats as an unwelcome extra. As an extension agent, I've received increasing numbers of fungus gnat specimens from consumers for identification. It's unfortunate that the greenhouse industry is the likely source.

Fungus gnat management isn't difficult; it just takes a deliberate effort to follow a comprehensive strategy or integrated pest management approach. The following suggestions should help.

First, be sure to correctly identify. Fungus gnats occur in four distinct stages. The most recognizable are adults—small, black flies with long legs and antennae. They can be seen quickly running across surfaces or weakly flying. Larvae can be observed in infested roots and the lower parts of plant stems. They're small, less than a quarter-inch long, wormlike, and clear in color with black heads.

Next, make it tough on them: Dry out and clean up! Moist greenhouse conditions are favorable for fungus gnat growth and reproduction. Your first line of defense is to minimize breeding areas through modifications in irrigation practices, sanitation, and greenhouse design. The most severe outbreaks are usually associated with unsanitary, wet greenhouses. Clean up all crop debris, weeds, and other sources of organic debris. Reduce overhead irrigation volume and leaching (but be sure to monitor media EC to avoid high fertilizer salt levels). Consider ways to improve drainage within production facilities; however, take measures to avoid potential environmental contamination. Increase airflow with HAF fans and wider crop spacing. Remember that crowded, over-watered crops result in wet conditions, and wet production facilities not only favor fungus gnats, but also infectious diseases.

There are many labeled pesticides for fungus gnat control. Before treating, however, be sure to monitor for adults and larvae. Adults are attracted to yellow sticky cards. Place them horizontally at the soil surface, one card per thousand square feet of production area. Place cards both at crop level and beneath benches. Make weekly counts to determine the locations of problems and population trends. Larvae can be detected by placing half-inch thick potato slices, raw side down, on soil surfaces. Some researchers recommend pressing pieces into the media so that the topside is flush with the media. If adult and larval stages are not present, treatment is not necessary.

Be sure to also inspect any crops entering your facilities. Check plugs, rooted cuttings, and prefinished plants. Don't forget Oasis cubes. Fungus gnats can infest these too.

When battling serious infestations, treat infested crops and greenhouse floors with drench applications of registered insecticides. Soil applications are targeted to control larvae. Recent research from Michigan State University showed excellent larval control in containers with Duragard, Marathon, Oxamyl, and Dycarb. Other labeled products include Knox-Out and the insect growth regulators Adept, Azatin, Citation, Enstar II, and Precision.

After treating for larval stages, monitor crop media and soil below benches with potato slices to see if your treatment was effective. If needed, use labeled aerosol insecticide formulations to clean up any lingering adults. Don't treat greenhouse crops or floors unless the label allows.

Biological controls can be effective, safe, and environmentally benign. Gnatrol contains bacteria that release toxins that kill fungus gnat larvae when ingested. Apply as a drench early in the crop cycle. If larval populations are high, repeat applications will be necessary every few days until numbers decrease. X-Gnat contains nematodes that attack and kill fungus gnat larvae. Apply as a drench early in the crop cycle. Nematodes are sensitive to drying and ultraviolet light—be sure to irrigate before applying. Applications during evening or cloudy weather will help ensure efficacy. Finally, the predaceous mite *Hypoaspis miles* may give excellent control. This species is adapted to similar soil conditions as fungus gnats. Hypoaspis can persist and become established in greenhouses.

When mites arrive, check them with a hand lens or magnifying glass—they must be alive to work! Promptly apply the recommended rate starting early in the crop cycle. Pesticides with miticidal properties will reduce effectiveness.

Pest Control, February 1997.

Stop Fungus Gnats from Spoiling Easter

Jim Willmott

Fungus gnats are a threat to profitability—in more ways than one! Most growers are aware of potential crop damage, but a greater concern is consumer objection to gnats brought indoors on infested plants. Lilies, bulbs, and flowering pot crops are a delight for consumers longing for spring, but not when accompanied by a cloud of gnats. With about two months left until Easter, you still have time to get the upper hand on gnats.

Management Begins with Monitoring

Fungus gnats are now present in different stages of development and in various locations. Management begins with monitoring production areas. Start by determining severity of infestation and which life stages are present—adults and/or larvae? Serious infestations will require different control tactics.

Several types of fungus gnats and related insects can infest greenhouses. Dark-winged fungus gnats, including *Bradysia coprophila* and *Bradysia impatiens*, are the most troubling because larvae feed on and damage roots. Research also indicates that they spread common pathogens, which cause root rot diseases.

Fungus gnat development includes four distinct stages beginning with eggs, followed by larvae, pupae, and adults. Larvae are clear to white with black heads. When fully grown, they are about a quarter-inch long and are active within an inch of the surface of potting media or soil. Larvae feed on algae, decaying plant material, and living plant roots. A single generation, from egg to adult, requires three to five weeks depending on soil temperatures. Since females can lay up to two hundred eggs a week, populations can build quickly. You need to control fungus gnats before temperatures rise in late March and early April.

Begin monitoring for adults and larvae now. Adults are easy to detect with yellow sticky cards. Place at least one card per thousand square feet, at about the same height as crops. It's also useful to place traps beneath benches since most difficult infestations originate from soil and plant debris, including potting media. Check traps once or twice weekly. Frequent counting helps to delineate locations that may need more intense management.

Larvae are often present, but not obvious—especially on crops where root feeding is minimal. Monitor by placing potato slices on media or soil beneath benches. Slices should be approximately one inch in diameter and a quarter-inch thick. Check slices frequently. Larvae may be difficult to see at first, but their black head capsules and movement are noticeable.

Chemical Controls

Heavy infestations require immediate action. Begin by targeting larval stages with an insecticide labeled for drenching on infested crops. Duraguard and Knox-Out are organo-phosphate products with good soil activity. Many insect growth regulators are also labeled for fungus gnats, including Adept, Azatin, Citation, Enstar II, and Precision.

When drenching, apply with enough water to move insecticide about an inch into the potting mix. Infested soil beneath benches should also be treated, but first be sure

This sticky card is too high—make sure they're placed at crop level.

this is allowed on the product label. Monitor treatment effectiveness with potato slices.

Duraguard and Knox-Out may give some control of adults. Because IGRs don't control adults, high populations will need treatment. Consider aerosols such as PT 1300 Orthene or Attain TR alone or in combination as a "space mix." Spray applications with either Decathlon or Talstar are effective, but thorough coverage is critical. The combined approach of targeting larvae and adults should clean up infestations.

Biological Controls

For less serious infestations, biological controls can be effective. Gnatrol contains bacteria that have toxins, which kill fungus gnat larvae when ingested. Apply as a drench early in the crop cycle. For high populations, repeat applications every few days until numbers decrease. X-Gnat and

Scanmask contain nematodes that attack and kill fungus gnat larvae. Apply early in the crop cycle as a drench. Nematodes are sensitive to drying and ultraviolet light, so irrigate before applying. Applications made during evening or cloudy weather will help ensure efficacy.

The predaceous mite *Hypoaspis miles* has provided excellent control. This species is adapted to similar soil conditions as fungus gnats. Hypoaspis can persist and become established in greenhouses. When mites arrive, check them with a hand lens or magnifying glass to make sure they are alive. Promptly apply the recommended rate starting early in the crop cycle. Pesticides with miticidal properties will reduce *Hypoaspis's* effectiveness.

Remember that as Easter approaches, warmer weather will favor rapid buildup of fungus gnats and other pests. As bulbs and other pot crops begin to enter production areas, step up your monitoring efforts. Incoming crops may have been infested prior to cold storage.

Don't overlook underlying reasons for chronic, reoccurring fungus gnat problems: Minimize wet areas and accumulation of crop debris. Good water management and sanitation are also essential for preventing future fungus gnat troubles.

Note: Inclusion or omission of products does not constitute endorsement by Rutgers University Cooperative Extension. Pesticide applicators are responsible for ensuring that product use complies with regulations and the product label.

Pest Control, February 1998.

Mealybugs

Tough New Mealybug Threatens Southern Crops

Chris Beytes

"It's not in the mainland yet—that's the good news," says entomologist Lance Osborne. "But it's going to be here—that's the bad news," he warns. Lance, of the University of Florida's Central Florida Research and Education Center, Apopka, is speaking about the recently identified hibiscus mealybug that's been devastating a wide range of plant life from ornamentals to shrubs to hundred-year-old trees in the Caribbean islands south of Florida. While it hasn't officially been reported in the state, at least one well-traveled industry consultant says he found hibiscus mealybug on *Mandevilla* at a couple of sites.

Hibiscus mealybug, *Maconellicoccus hirsutus,* also known as pink mealybug, was first identified in Grenada in the early 1990s. It has made its way through the tropical islands of the Caribbean and was found in St. Thomas, the Virgin Islands, in May. Being wind-borne, it is expected to migrate quickly to nearby Puerto Rico, and Lance says a tropical storm or hurricane could easily carry it to Florida's mainland. It can even move from one place to another on clothing, making its spread extremely difficult to control.

The pest infests more than 200 different plants, including *Aralia, Croton, Ficus,* chrysanthemums, hibiscus, and *Dracaena*—"a whole array of things we import," says Lance.

The wide range of hosts and the sheer numbers of the pest are part of the problem; the serious damage it can cause is the other. Hibiscus mealybug is one of the few mealybug species that exudes a toxin that gnarls and curls the new foliage of its host plants. It also shortens plant internodes and stunts plant growth, causing "bunchy top," a characteristic of infested plants.

Lance says chemical controls have proven to be fairly ineffective against hibiscus mealybug, but he says Dursban has shown some potential. In Hawaii (the only state where the pest is found), a natural parasite keeps the mealybug under control. Lance is working with the University of Florida and state and federal officials to build a "safety net" in Florida of two natural parasites by identifying mealybugs that predators can feed on until hibiscus mealybugs come into the state.

What can growers do? First, learn to identify hibiscus mealybug. Unlike citrus or longtail mealybugs, which have fringed bodies, hibiscus mealybug is smooth and waxy, with just two short anal filaments. Second, when you squash a hibiscus mealybug, its body fluid is dark red instead of the clear fluid typical of other mealybugs. Third, look for the characteristic leaf curling and distorting it causes. If you're importing plants or cuttings, inspect them closely. A registered dip with Mavrik seems to work and has even shown to aid plant rooting, Lance says.

For more information, contact Lance Osborne at the Central Florida Research and Education Center, Apopka, Tel: (407) 884-2034. Lance says they'll soon have pictures of the pest on their Web site at www.mrec.ifas.ufl.edu.

GrowerTalks in Brief, September 1997.

Mites

Cyclamen Mite Is Back

After years of few infestations, cyclamen mite has resurged. The number of cases diagnosed at Long Island Horticultural Research Laboratory has increased, Daniel Gilrein, entomologist, Cornell Cooperative Extension of Suffolk County, Riverhead, New York, reports in *Greenhouse IPM Update*. First found in the U.S. on verbena in Illinois in 1883, cyclamen mite, *Phytonemus pallidus*, is now a pest in much of the world and can be a problem in the greenhouse year-round.

Hosts include many plants other than cyclamen: African violets, begonias, chrysanthemums, *Exacum*, fuchsias, gerberas, impatiens, English ivy, geraniums, kalanchoes, petunias, snapdragons, strawberries, verbenas, a variety of tropical foliage plants, and many others. These mites can be spread on propagative material; on birds, insects, and equipment; and between adjacent plants with overlapping foliage.

Symptoms often mimic a disease, including stunted, distorted, brittle, and sometimes thickened or discolored foliage. Leaves may not open or may have tightly rolled margins. Plants can also appear streaked or blotched, be denser than usual, or have fewer flowers than normal. Broad mite can cause similar damage but is distinguishable by its egg, which has a "hobnailed" appearance.

At only one-hundredth of an inch long, cyclamen mites are virtually impossible to see without a 20X hand lens. They may be difficult to detect even under a stereoscope, as they often hide in buds or folded leaves. Look for them in the newest growth, within leaves or among leaf hairs.

Females are a translucent yellow or orange and are more long than round. Males are light brown with a claw on each back leg and are smaller than females. Females usually outnumber males and can reproduce without mating.

Eggs are relatively large (about half the size of the female), and females lay one to three per day. They are oval, smooth, and nearly transparent. Larvae hatch from eggs after three to seven days, then feed for one to four days before transforming to pupae. Adults emerge after resting for two to seven days. Each generation lasts one to three weeks, depending on temperature.

Controlling a cyclamen mite infestation can be as difficult as finding the mites. If you find an infestation, discard infested plants and handle crops carefully to avoid transporting mites to clean areas. To limit mites' movement, space plants to avoid touching. Research shows that immersing infested plants in hot (110°F) water for fifteen to thirty minutes is effective, but it may not be practical. It may not affect young plants, but it may injure cyclamen in bud and blooming stages.

Information on biological control of cyclamen mite is limited; however, predatory mites including *Neoseiulus californicus* and *N. cucumeris* will feed on cyclamen mite.

Chemical control is possible but difficult, as mites live in folded leaves or buds. Endosulfan (Thiodan) and dicofol (Kelthane) are labeled for control of cyclamen mite on ornamentals. It's important to repeat high-gallon applications to maximize penetration into buds and other areas and to expose mites that may have been missed with a previous treatment. Research on California strawberries has revealed that cyclamen mite may be resistant to endosulfan where it has been used for many years and that dicofol has been less effective than in the past.

Culture Notes, January 1998.

Summer Heat Won't Slow Down Spider Mites

Jim Willmott

As we move into summer, don't let the heat slow down your pest management efforts. Spider mites love warmth and many greenhouse and field-grown ornamental crops. At risk are fall garden plants, perennial plugs, cut flowers, and foliage plants. Spider mites are tiny but explosive—they often catch growers by surprise. Prevent crop loss now by implementing a comprehensive, integrated management strategy.

The Basics

First, mites aren't insects. They're actually more closely related to spiders. Several types are economically significant, including spider mites, false spider mites, and broad, cyclamen, and eriophyid mites. While all may be troublesome, the two-spotted spider mite, *Tetranychus urticae,* is most common. It will attack more than three hundred species of plants by piercing plant cells and sucking up the contents. At first, only a few tiny spots or stipples may be evident, but soon plants appear bronze or white with spiderlike webbing.

Two-spotted spider mites have a simple life cycle that begins with eggs, then passes through two nymphal stages before reaching the adult stage. During hot summer conditions, a complete generation takes as few as five days. With females laying as many as a hundred eggs, populations can build in a very short time. Mites move through crops quickly, especially if spacing is close. They can also move through air currents or by crop handling.

Preventing Infestation

Preventative management begins with sanitation. Before production, eliminate all weeds inside and adjacent to greenhouses and field production areas. Consider planting low-maintenance turf grass to reduce weeds, spider mites, and other pests. Remove infested "pet plants" from greenhouses, and inspect incoming propagation stock.

Finally, establish a formalized monitoring or scouting program for early mite detection and problem delineation. Because mites are so small, magnification is necessary. Provide scouts with hand lenses or hands-free magnifiers that attach to headbands (Optivisor is one brand). Plants can also be sampled by shaking them over white paper or cloth.

Control Options

Once infestations are detected, you can apply miticides, but avoid unnecessary treatments, which are costly and increase the chances for resistance. Be sure that products are labeled for application to infested crops and that you follow all precautions before treating. While many miticides are labeled, there are resistance problems. Before setting up a program, check with extension specialists for what's best in your area.

Generally, you should repeat applications at least two times at five- to seven-day intervals. Rotate to a new chemical class after three to four weeks. Most miticides are contact pesticides with no systemic activity. Because mites primarily feed on lower leaf surfaces, spray coverage is critical.

Several miticides are labeled for use in the U.S. In recent years, Avid has been a standard providing excellent control. Its translaminar activity helps by allowing movement from upper to lower leaf surfaces where mites feed.

More recently, Sanmite was introduced with positive results in university trials. While it's relatively slow to act, it has good residual activity, controlling mites for five or more weeks.

Many pyrethroids—including Attain (aerosol), Decathalon, Mavrik, Talstar, Tame, and Topcide—may be effective where populations aren't resistant. Pentac and Kelthane are in the chlorinated hydrocarbon class and are among the oldest pesticides. They're effective, but Pentac is no longer being manufactured and is no longer registered or for sale. Existing supplies may be used. Knox-Out, Sulfotepp, DDVP, and Dithio are organophosphates that may suppress populations.

Finally, insecticidal soaps and horticultural oils can be very effective because there is no resistance to them. Use caution when using soaps and oils, as some plants may be injured. Use the lowest possible concentration—never more than a 2% spray volume—and avoid treating plants under heat or moisture stress. Both soaps and oils can increase the effectiveness of other miticides when tank mixed.

Two-spotted spider mite damage on upper leaf surfaces of *Celosia plumosa*.

Biological control is also an effective management alternative, but it requires knowledge of the various mite predators. Several species of predatory mites are commercially available, and each has unique qualities that affect its performance. The most commonly used is *Phytoseiulus persimilis*, which aggressively pursues and consumes up to twenty young mites per day. *Mesoseiulus longipes* survives lower humidity and higher temperatures. *Neoseiulus californicus* is slower acting but can survive longer periods without prey. Releasing a mix of different species offers adaptability to changing environmental conditions and prey populations.

Success with predatory mites requires early spider mite detection and prompt release of predators onto infested plants. Because most miticides and insecticides are toxic to predators, avoid using them both before and after releasing predators. You can apply insecticidal soaps or horticultural oils before releases, as they have no residual activity after drying. Other compatible pesticides include Marathon, insect growth regulators, *Bacillus thuringiensis,* and entomopathogenic nematodes.

Finally, cultural management can reduce mite problems. Begin by avoiding overfertilization with nitrogen. The resulting soft tissue is attractive and nourishing not only to mites, but also to many sucking insect pests and some pathogens. Grow resistant plants and cultivars and avoid those that are frequently troubled. While there is limited information on resistant plants, document those that you find infested and keep records for the future.

Pest Control, July 1998.

Nematodes

Soil Solarization Approved for Nematode Treatment

California's commercial nurseries now have an approved alternative to methyl bromide and steam sterilization of soils for nematode control: solarization, which is using the sun to heat soil to a temperature high enough to kill anything living in it.

Researchers at the University of California have been studying solarization for twenty years. Scientists at UC's Kearney Agricultural Center perfected the solarization techniques that have been approved by the California Department of Food and Agriculture for use by the state's commercial nurseries.

The approved system requires nursery owners to put soil onto a layer of plastic or a disinfected concrete pad. Soil must be moistened to near field capacity. The soil then gets covered with a layer of clear plastic. A second layer is suspended over the first layer, in effect creating a double-poly roof. The temperature of the soil at the bottom of the pile must climb to 158°F or higher for a minimum of thirty minutes, which will reduce nematode populations to "undetectable" levels, while killing weed seeds and soilborne pathogens.

According to the researchers, solarization should only cost growers about $500 an acre, compared with $1,200 to $1,400 per acre for methyl bromide treatments. The system is expected to be popular in warmer areas of the state, such as the San Joaquin Valley.

Culture Notes, August 1999.

Slugs and Snails

Slug 'Em

Tom Thomson

Slugs and snails are major pests in greenhouses and nurseries. They're in the mollusk family, which also includes clams, oysters, and octopi. Snails have a hard outer shell, while slugs have soft, slimy bodies without shells. Both pests damage plants by eating them.

Snail and slug damage is often misdiagnosed as insect damage. Often, growers try to treat with insecticides, but these products have little effect on mollusks since their metabolisms are completely different than insects'.

Snails and slugs spend most of their time hiding in dark, damp places, such as under weeds, leaves, groundcovers, rocks, and woodpiles. They come out at night or early in the morning to feed and can eat the equivalent of their body weight in a single feeding. Both can be detected by the slimy trails they leave behind. Due to their strong sense of smell, they'll travel a substantial distance to find a plant of choice—petunias, marigolds, artichokes, citrus trees, hostas, and orchids are some of their favorites.

Control Methods

Sanitation

Cleanliness and good maintenance are the first methods of control. Eliminate the areas where they spend their daylight hours. Clean up weeds and undesirable plants in and around your crops. Water early in the day to eliminate moist areas where they congregate. Clean up all wood, rocks, and other areas where they hide.

Biocontrol

In some areas, decollate snails can be used for biological control. They live on the eggs of snails and slugs and eat rotting plant materials, but they won't eat live plants. Over a period of six months or longer, they can eliminate the population of brown garden snails and reduce the population of slugs. Release them at the rate of one thousand per acre.

Chemical control

Everyone has heard about beer, salt, and other homegrown snail and slug controls. These aren't practical under production conditions, however.

Metaldehyde baits, first introduced in the 1930s, are the most commonly used control. Metaldehyde is usually formulated as a meal bait or pellet, from 2.0 to 7.5% active ingredient and sold under various trade names; one product is Deadline, a 4.0% metaldehyde gel formulation. Metaldehyde kills through both ingestion and contact. If a slug or snail crawls across the metaldehyde bait, the metaldehyde is taken up into mollusk's system, causing death. Some baits have a combination of metaldehyde and carbaryl.

Metaldehyde baits are registered for use on ornamentals, tree fruits, greenhouse ornamentals, berry crops, vegetables, and grasses grown for seed and turf areas. Don't allow the bait to contact any edible portions of any food or feed crop. Phytotoxicity has been observed in some ornamentals, including daylilies and clematis. Apply when plant damage first appears, and don't apply to a dry soil. Evening applications are preferred. For heavy infestations, a second application may be required in seven to ten days. Metaldehyde baits are extremely toxic to pets, so keep them out of treated areas.

Mesurol (methiocarb) is a sprayable molluscicide/insecticide that's registered in the U.S. on greenhouse and field ornamentals. It's applied as a foliar spray up to four times per season. Mesurol controls the snails and slugs when they feed on the treated foliage. Once ingested, feeding stops. Mesurol also controls aphids and mites. It's not for use on food or forage crops.

The newest bait to hit the market is Sluggo (iron phosphate). Developed in Germany, it's registered in the U.S. for use on ornamentals, greenhouse vegetables and ornamentals, citrus, fruit crops, field crops, vegetables, and grasses grown for seed, turf, and landscape areas. The active ingredient is incorporated into a pasta-type product. Sluggo can be used up to the day of harvest on food or forage crops. One of the biggest advantages of Sluggo is that it's safe for pets and animals. It has no effect on earthworms, birds, insects, or other nontarget species.

Iron phosphate, the active ingredient, occurs naturally in the soil and is used in medicine to correct iron deficiency. It breaks down in the soil into iron and phosphate, both of which are used by plants as fertilizer. The inert ingredients are all biodegradable food additives.

Unlike metaldehyde baits, Sluggo must be ingested to be effective; it doesn't kill on contact. Once the heavy metal (iron) is ingested, it becomes toxic to the pest. The snails or slugs stop feeding immediately and go back to their nesting area to die. You won't see unsightly piles of dead snails or slugs when using iron phosphate bait.

Apply Sluggo around the plants to be protected. It may even be scattered over the top of ornamentals and vegetables. Evening is the best time to apply the bait. Reapply as bait is consumed.

Pest Management, August 2000.

Thrips

You Can Control Thrips Biologically

Stanton Gill

Thrips continue to be one of the most difficult pests to keep at manageable levels in commercial greenhouses. Several factors make thrips difficult. Adult female thrips use a sawlike ovipositor to pierce plant tissue and deposit eggs in protected locations on the plant. For most species, eggs are inserted into leaf tissue and are difficult to detect. Thrips have a high reproductive capacity, with the number of eggs deposited by females varying from thirty to three hundred, depending on species. Thrips can reproduce sexually (male and female mate) or asexually (female is self-fertile). The larval stages of several of the major species attacking greenhouse plants feed in small, confined areas of plant material where contact with pesticides is difficult. Thrips' life cycle is relatively short, creating problems for growers having to deal with multiple generations per growing season. In addition, thrips rapidly develop pesticide resistance, and several species vector viruses that render plants unsalable.

As part of the greenhouse IPM efforts of the University of Maryland Cooperative Extension, we're continually looking for effective biological control options for commercial greenhouses. In 1996, we evaluated several thrips control options in the

Use a fine mist (in this case, through a Dramm sprayer) to apply *Beauveria bassiana* because the material must come into contact with the pest.

field: two biological control agents (a predacious mite and an entomopathogenic fungus for thrips) and a relatively new insect growth regulator.

Preliminary trials in 1995 indicated that the entomopathogenic fungus *Beauveria bassiana* appeared to be a viable, affordable material with potential as part of a thrips management program. Also, two predacious mite species, *Amblyseius* (=*Neoseiulus*) *cucumeris* and *Amblyseius* (=*Neoseiulus*) *degenerans*, have shown promise when released early in a crop cycle before thrips populations build up. We've been using *N. cucumeris* in Maryland greenhouses involved with our TPM/IPM program over the past three years. We were also interested in investigating the efficacy of a new insect growth regulator, Precision, in controlling thrips. Insect growth regulators as a general rule have the least damaging effect on many beneficial organisms used in greenhouse biological control.

Predacious Mites for Thrips Control

Amblyseius (=*Neoseiulus*) *cucumeris* is a predacious mite that feeds on young thrips. The eggs of *N. cucumeris* are slightly larger than two-spotted spider mite eggs and are usually laid on leaf undersides. There are two nymphal stages. Adult mites are very mobile; females are slightly larger than males. Eggs become adults in six to nine days at 77°F. Mites develop best at temperatures of 80° to 85°F. Adult mites can live up to thirty days feed on an average of one larval thrips per day during that span. The mite can also survive on pollen from flowers.

The mite is most effective in feeding on first-instar thrips larvae, so they need to be introduced before the thrips population builds up. Mites will reproduce in the greenhouse unless applications of broad-spectrum pesticides are made. They are relatively tolerant of applications of insect growth regulators. Mites are most effective in greenhouses from March through October. During the short days of winter, mites go into a resting stage (diapause). Mites can be introduced into a greenhouse crop through a shaker dispenser; they're also available in sachets containing a breeding colony of mites. Sachets are spaced in the greenhouse, and the corner of the bag is removed to allow mites to move onto the crop. Don't place sachets where they'll be continually wet, or the grain carrier in the bag will ferment, killing the mites in the bag.

Neoseiulus degenerans is another predacious mite that feeds on thrips. The lifecycle is very similar to *N. cucumeris*. The mite feeds mostly on the first-instar larvae of thrips and can survive on pollen from flowers. This mite is a more voracious feeder than *N. cucumeris,* consuming four to five thrips per day. It is also generally more expensive than *N. cucumeris.* These mites can be used in winter months because this mite doesn't go into a resting stage (diapause).

Precision Insect Growth Regulator

The new insect growth regulator Precision (fenoxycarb), from Novartis, has an oral LD50 of greater than 5,000, a dermal LD50 of greater than 2,000 and a twelve-hour REI. The material prevents chitin formation in the insect, working on the larval stages of thrips. It's labeled for whiteflies, aphids, and thrips in the greenhouse. With most insect growth regulators, it's important to begin applications when populations are very low rather than using materials when you have an exploding population.

Entomopathogenic Fungi

Researchers are investigating several entomopathogenic fungi for control of greenhouse pests, including *Metarhizium* spp., *Paecilomyces fumosoreus,* and *Beauveria bassiana.* The infective unit in entomopathogenic fungi is the conidium, which may penetrate the insect cuticle from a combination of mechanical pressure by the germ tube and enzymatic degradation of the cuticle. Once through the insect cuticle, the fungus proliferates as hyphal bodies. The fungus spreads through the insect body, and the insect normally dies within two to fourteen days after spore application.

The fungus *B. bassiana* has shown to be highly pathogenic to aphids, whiteflies, leaf-feeding caterpillars, flea beetle, lygus bugs, and spider mites. In greenhouse trials conducted by the University Maryland Cooperative Extension, *B. bassiana* has performed well in controlling greenhouse whitefly

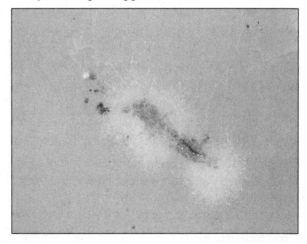

Western flower thrips infected with *B. bassiana.*

Chart 1. Treatments Used in the Greenhouses

Treatments and greenhouse number	Frequency (once a week application)	Comments
Greenhouse 1 *Beauveria bassiana*	2 lbs. A.I./100 gal. per acre	Foliar application
Greenhouse 2 *Beauveria bassiana*	2 lbs. A.I./100 gal. per acre	First application applied as soil drench, following applications made as foliar application.
Greenhouse 3 *Beauveria bassiana*	1 lbs. A.I./100 gal. per acre	Foliar application
Greenhouse 4 Precision (fenoxycarb) and Cell-U-Wet (170 g. in 20 L of water)	1 prepackaged pack in 25 gal. of water	Foliar application
Greenhouse 5 Precision (fenoxycarb)	1 prepackaged pack in 25 gal. of water	Foliar application
Greenhouse 6 *Neoseiulus cucumeris*	60 sachets placed in greenhouse	Sachets placed in greenhouse on March 1. A second set of sachets was supposed to be placed for the second crop, but they weren't ordered on time.
Greenhouse 7 Control greenhouse		No treatments made

and chrysanthemum aphid. Repeated applications (three to five) at five-day intervals are necessary for best control. The material must be applied using a fine mist as it must make contact with the pest. Two commercial manufacturers have received EPA labels for products to control greenhouse pests. One is BotaniGard by Mycotech, and the other is Naturalis-O by Troy Bioscience.

July 1997.

Thrips Tips

John Sanderson

The western flower thrips continues to top the list of the most serious greenhouse pests for most growers. Feeding by these tiny insects causes plant cells to collapse, resulting in scarred patches if the damage is to open foliage or petals and distorted leaves or flowers if the damage was done while the thrips were feeding within buds and terminals.

The thrips' feeding can also transmit one of the two incurable tospoviruses: impatiens necrotic spot virus (most common) and tomato spotted wilt virus. Both cause severe damage to greenhouse plants. Both thrips and the viruses have a very wide host range, including flowers, vegetables, and many weeds.

Here are some suggestions for managing these pests:

Clean house. Thrips control should start at the end of the previous crop or season. Eliminate all sources of thrips at the end of each crop or growing season to avoid harboring a small population ready to infest an incoming crop.

Keep weeds eliminated for the crop's duration. They may harbor thrips and/or viruses (some plants don't show viral symptoms). Stock plants and flowering pot plant crops such as cyclamen or holiday cactus might also serve to carry thrips over from the fall to the following bedding plant season, as might weeds left under the benches or just outside the greenhouse.

Check new plants. Inspect incoming plant material for signs of thrips damage. Insist on good thrips control from your plant suppliers. Baskets of infested cutting crops hanging over seedlings can spread a thrips infestation and possible viruses to the bedding plants below.

Cut the flowers. Remove unnecessary flower buds and blossoms from plants, if possible; place gently into plastic bags and seal before disposal. Buds and open flowers provide an ideal habitat for thrips, and their numbers can greatly escalate within them. Some growers have seen a noticeable drop in thrips numbers after unnecessary flowers have been removed and discarded.

Screen them out. If thrips may be entering the greenhouse from outside sources, consider screening the vents. The National Greenhouse Manufacturers Association (NGMA) has compiled useful research data in its publication "Insect Screen Installation Considerations for Greenhouse Operators." It can be obtained from the NGMA office, (800) 792-6462.

Stick 'em up. Use yellow or blue sticky card traps to monitor for western flower thrips. Count and change cards weekly, noting upward population trends that signal the need for treatment. Use three-by-five-inch sticky cards spaced every thousand square feet and positioned vertically just above the top of the plant canopy. Some growers who've placed cards near doors or vents have noted that thrips are more numerous on the sides of the cards that were facing a source of thrips. So the sticky cards may also be useful to detect when and where the thrips are originating.

Flowers can also be checked for thrips by tapping a blossom over a white tray or sheet of paper, although it's usually more efficient to use sticky cards to monitor trends in thrips levels. Damage to open foliage will normally be on the upper leaf surface and look like small scarred patches with irregular outlines that usually have tiny black fecal specks of thrips.

Conserve, Avid (though not yet labeled for thrips), Mesurol, DuraGuard, Dycarb (Turcam), Orthene (TT&O and PT1300), Azatin*, Neemazad*, Knox-Out, Precision*, Talstar, Decathalon, Mavrik, Topcide, Preclude*, DuraPlex TR, Carzol (in Indiana, North Carolina, Ohio, South Carolina, and Utah only) may be used for thrips control or suppression on floral crops (see labels for crops and directions). Starred (*) materials are insect growth regulators and will not kill the adult stage. Growers have reported good results with Conserve, Mesurol, and Avid.

Because thrips are or can become tolerant or resistant to most insecticides and are difficult to control at best, emphasis must be placed on sanitation and other nonchemical preventive measures. Also, to preserve the effectiveness of new insecticides such as Conserve, save them for when you really must have good control.

Follow label directions for each insecticide used, but often a five-day treatment interval for two to three applications is needed to have an effect on thrips that hatch from eggs or emerge from pupae soon after the initial insecticide application.

Use an application technique (sprayer, aerosol, fogger, etc.) that will give good pesticide coverage with tiny droplets. The insecticide should penetrate into thrips' hiding places such as buds and terminals, so tiny spray droplets are important. Pump-up sprayers may not be adequate.

BotaniGard and Naturalis-O are two formulations of an insect-killing fungus (*Beauveria bassiana*) that some growers have found effective for thrips. Three to five applications at three- to five-day intervals have often kept low to moderate infestations in check. The addition of Azatin may increase effectiveness.

Pest Control, July 1999.

Deter Thrips with Reflective Fabric

Like a high-tech scarecrow, aluminized shade fabric on greenhouses can deter western flower thrips from entering greenhouses by as much as 55%, according to Colorado State University researchers. Levels of thrips entering the greenhouse decreased more as the fabric was placed closer to the entrance, regardless of surface area covered.

In a series of experiments, researchers tested the effectiveness of woven aluminum fabric and reflective tape in deterring thrips' entry into greenhouses. Using Aluminet, a slitted aluminum fabric intended to provide 50% shade, and a reflective tape, researchers tested different patterns including a solid area of fabric, a checker board pattern, and a border trim around the outside perimeter. Fabric was attached to six wood cages (two for each treatment) with yellow sticky cards inside. Researchers covered the cages with plastic, cut an entrance in the plastic and placed them outside. Researchers also included two control cages with only plastic.

When the aluminum fabric actually covered the hole in the cage, thrips entry was most reduced, indicating that the fabric may work as a screen as well as a reflector. Similarly, fabric or reflective tape placed closest to the entrance and monitoring area were more effective in deterring thrips. Researchers found no evidence that applying the fabric in patterns was more effective than other

Thrips Captured at a Colorado Greenhouse

	Number of thrips per trap*	
Treatment type	Week 1	Week 2
Control (No Aluminet)	31.3a	39.3a
Hanging strip	17.8b	19.3b
Base strip	12.5b	13.0b
Base and hanging strips	13.8b	17.3b

*Mean from four traps. Numbers followed by the same letter within the same week weren't significantly different.

applications. However, one experiment did reveal that the reflective tape may be more effective than the Aluminet in deterring thrips.

Also, researchers found that the amount of surface area may not be as important as the reflective capacity of the material. Applying the tape around the inside of the cage hole and using a pattern with Aluminet that covered twice the surface area did not significantly reduce thrips' entrance.

In addition, researchers tested the aluminum fabric on west roof vents on a hydroponic tomato greenhouse in August. They hung the fabric from the top of an open vent, put a base strip along the bottom of the open vent, and used a pattern with both a hanging strip and a base strip. As a control, one vent was left untreated. They hung yellow sticky cards in front of the vents and counted them twice at one-week intervals.

In the greenhouse trial, researchers found no significant differences between treatments, but all treatments differed from the control. This could be because the fabric directly blocked access to sticky traps and provided screening. Or it could be attributed to heat leaving the greenhouse through the vents, raising the fabric so that it was horizontal above the vent. This would make the treatment similar to the strip applied along the vent base. In addition, thrips entry wasn't reduced when the hanging and base strips were used together, suggesting that less material provides similar results. Proximity to the vent was most influential in deterring thrips' landing.

Culture Notes, January 1997.

Battling Thrips: Five Pesticides Put to the Test

Stanton A. Gill, Rondalyn Reeser, and Michael J. Raupp

The western flower thrips is one the most widespread, common, and difficult to control pests in commercial greenhouses nationwide and internationally. As a vector of tomato spotted wilt virus (TSWV) and impatiens necrotic spot virus (INSV), this insect pest has caused epidemic crop losses in greenhouses across the U.S.

To make matters worse, researchers have shown that western flower thrips has acquired resistance to many pesticides—but rotating chemical classes can help delay more serious resistance problems with this damaging insect.

To help growers learn how effective various commercial chemicals are at controlling these thrips, we conducted trials in a Maryland commercial greenhouse to evaluate several materials on garden mums. While two of the products tested aren't labeled for thrips control, our results should help you in choosing products in your battle against these tough insects.

Small but mighty. The western flower thrips is one of the toughest insect pests that growers face and for two main reasons: They're small, so they're difficult to kill, and they can spread viruses, leading to severe crop losses.

The Products

One product we tested is Conserve from Dow Elanco. It's a new class of chemical called spinosyns. Conserve is derived from a natural-occurring bacterium. The toxin produced by the bacterium kills insects through contact and ingestion.

Two strains of *Beauveria bassiana* were included in the trial: Naturalis-O from Troy Bioscience and BotaniGard from Mycotech Corporation. *Beauveria bassiana* is a fungus that penetrates the insect's body and kills it either with a toxin or as a result of fungal growth inside the insect. Death takes from two to fourteen days after application.

We also tested ZeroTol, a new material from BioSafe Systems. Its active ingredients are peroxyacetic acid and hydrogen dioxide. This material is presently labeled for use in the greenhouse as an algaecide and fungicide, but it's been reported to have suppressant activity against some pests, a theory we wanted to test.

The fourth material used in the trial was Sanmite, manufactured by BASF. Its active ingredient is pyriban, and it's formulated as wettable powder. Sanmite is presently labeled for spider mite control in greenhouses, but thrips control has been reported by some growers, so we thought it was worth a look.

Setting up the Test

Because this trial was conducted in a production greenhouse, it was difficult to conduct true replication. We treated fourteen different varieties of garden mums grown in single greenhouse bays, ranging in size from 5,900 sq. ft. (1,700 plants) to 7,500 sq. ft. (2,000 plants), with five untreated control plants and five treated control plants designated in each greenhouse. The bays are separated by walls, providing fairly good isolation for each test. Control plants were removed from each bay before each treatment and returned twelve hours later. Table 1 shows the pesticides, rates, and treatment dates that we used.

We didn't introduce thrips into the greenhouse bays, but allowed them to build up naturally. On June 7, prior to starting the study, four yellow sticky cards were placed on stakes slightly above the plant canopy in each bay. We recorded the total number of thrips nymphs collected every seven days in each greenhouse.

Table 1. Treatments Applied to Each Greenhouse

Treatment	Area treated	Rate	Treatment dates (1997)
Conserve	5,900 sq. ft.	3.9 oz./30 gal.	6/23, 6/30, 7/7
Conserve	5,900 sq. ft.	7.8 oz./30 gal.	6/23, 6/30, 7/7
ZeroTol	5,900 sq. ft.	37.2 oz./30 gal.	6/23, 6/26, 7/7, 7/10, 7/21
Sanmite	5,900 sq. ft.	.8 oz./20 gal.	7/13, 7/21
Naturalis-O	7,500 sq. ft.	40 oz./40 gal.	6/18, 6/26, 7/8, 7/16, 7/22
BotaniGard	7,500 sq. ft.	13.6 oz./40 gal.	6/18, 6/26, 7/8, 7/16, 7/22

All treatments were made using a high volume Agrotec sprayer run at 300 psi. All applications were made to the point of runoff and thoroughly coated the undersides and tops of leaves.

Thrips samples were taken once a week by an IPM scout on the five control plants and five treated plants using a tap test to dislodge hidden thrips. Whole-house counts of thrips that were captured on the yellow sticky cards were also recorded weekly.

When we started the trial, the mum plants where the Conserve was to be used had thrips larvae present at densities of three to four thrips per plant. In the bays receiving two rates of Conserve, a switch to Sanmite was made after three treatments, in accord with the proposed label for Conserve that will state that after three applications you should rotate to another class of chemical.

Treatment with BotaniGard, Naturalis-O, and ZeroTol were started once thrips were found on the yellow sticky cards and were continued at weekly intervals. Midway through the trial, we increased the frequency of the ZeroTol applications, with two applications at four-day intervals. When ZeroTol failed to reduce thrips populations, the grower treated the area with Sanmite.

Because the greenhouse structures were unscreened, continual populations of adult thrips were recorded on the weekly yellow sticky card counts. No significant difference was noted in adults thrips caught on sticky cards in the greenhouses from June 17 through July 22. The sticky card counts taken on July 29 did have significantly higher thrips populations in the Naturalis-O- and BotaniGard-treated greenhouses compared with the Sanmite- and Conserve-treated houses.

What Worked and What Didn't?

ZeroTol applied weekly didn't give significant control even when we increased the application frequency. But remember, it's not labeled for insect control, so this shouldn't be surprising. Sanmite applied in the ZeroTol-treated greenhouse gave significant control of thrips on July 23 and July 29.

Conserve at the highest rate performed well, providing significant control on July 3. Populations remained low throughout the study. Conserve at the lower rate had no detectable effect on thrips. The Sanmite treatments that followed the Conserve treatments reduced the thrips populations significantly, below the control plants on July 15 in the greenhouse.

BotaniGard gave significant control of thrips populations on June 24 and July 29. No additional pesticide applications were necessary—the plants were healthy and were later sold.

Interestingly, the plants in the Naturalis-O-treated greenhouse showed very little difference between the control and the treated plants; however, the number of thrips on both the controls and the treated plants was very low. It's possible that infected thrips spread the *Beauveria bassiana* from the treated plants to the controls, resulting in the low thrips counts in both the BotaniGard- and Naturalis-O-treated greenhouses.

In conclusion, both the chemical treatments and the biological controls gave good treatment against western flower thrips when properly used and, in the case of Conserve, at the higher rates. If you're using Sanmite for mites, you'll get thrips control at the same time—a nice benefit. We also showed that, while very useful as an algaecide/fungicide, ZeroTol isn't effective against thrips. Your best bet is to use labeled products and adjust the rate by carefully monitoring insect populations—a key to every good IPM program.

Editor's note: The authors would like to thank Gary Mangum, owner, and Kathy Miller, head grower, at Bell Nursery for the use of their facilities to conduct this research trial. Partial support of this research was provided by the Maryland Agricultural Experiment Station, Maryland Cooperative Extension, and Environmental Stewardship Grant Program, and Environmental Stewardship Grant Program of the Environmental Protection Agency.

October 1998.

Biological Thrips Control on Bedding

Stanton Gill, Rondalyn Reeser, John Speaker, and Ethel Dutky

Because bedding plants mature quickly, it's been standard thinking that there just isn't enough time available to effectively use biological control of pests. Many growers feel that the crop time is simply too short to get biological controls established.

Maybe that's true, maybe it's not. With a few innovative application methods, we felt we might be able to make biological control work in a commercial greenhouse bedding plant operation. To test our theory, we decided to focus on just one major pest, western flower thrips (WFT). It's the predominant bedding plant pest in our state, causing plant damage and spreading impatiens necrotic spot virus (INSV).

In the spring of 1999, working closely with the staff at Bell Nursery, Burtonsville, Maryland, we treated and monitored three rotations of bedding plant crops from March through May. Greenhouses received an application of one of the following treatments through the three crop cycles:

- *Beauveria bassiana* (BotaniGard at 1 lb. per acre rate)
- *Amblyseius cucumeris* (12,000 mites released in 3,000 sq. ft. of growing area on weekly basis)
- *Beauveria bassiana* and the predacious mite Amblyseius cucumeris (BotaniGard at 1 lb. per acre rate and 12,000 mites released in 3,000 sq. ft. of growing area weekly)
- Control (no treatments)

To make it a real-world, practical demonstration rather than a controlled, replicated study, we treated entire greenhouses and maintained three control greenhouses. All were of the same size and had similar mixes of bedding plants.

Avoiding INSV

Before starting the experiment, we wanted to minimize the risk of INSV and other viruses. The grower started by making sure all weeds in the greenhouse were controlled, and his trained professional scout, John Speaker, made sure that all plugs and other starter material coming into the houses were virus free.

John also placed dwarf fava bean plants in the greenhouse to detect whether INSV was being actively transmitted by thrips in a greenhouse. Fava beans will show telltale dark-rimmed lesions on the foliage if virus-carrying thrips have fed. Adult thrips are monitored in the greenhouses using sticky cards placed at one card per thousand square feet of growing area.

Our objective was to kill larval stages of the thrips and prevent a flush of adults that would have the opportunity to lay eggs and feed. To accomplish this, biological control had to be applied early—in the first week that a crop was moved into the growing area.

Just to be safe, we guaranteed the grower that if the population of thrips reached an uncomfortable level (causing noticeable injury) or INSV was detected in the crop, we'd end the demonstration so he could use chemical treatments to control the thrips. In return, the grower agreed not to use any chemicals if we kept the thrips population low using the biologicals.

The Tim Allen Applicator

Since bedding plants are in the greenhouse for a short time (usually just two to six weeks), we knew the beneficial mites needed to be applied quickly and evenly. Based on an idea from Koppert Biological Systems, we bought a portable gas-powered leaf blower, drilled a hole in the blower tube, and attached a plastic hose connector with epoxy. A ball valve "trigger" was screwed into the hose connector and a standard two-liter soda bottle was screwed onto the trigger. When the blower was turned on, the ball valve was slowly opened to measure out the mites and grain carrier into the air blowing through the leaf blower tube. Tests showed that our blower would disperse the mites eight to nine feet, with an 80% survival rate for nymphs and close to 100% survival rate for eggs.

Leafblower retrofitted with bottle dispenser to apply the predacious mites. Tim Allen would be proud of this "more power" applicator.
Photo by Stanton Gill.

Starting the Test

The biggest cost in using small quantities of mites is shipping. To get around this, we ordered a large shipment of 50,000 mites every four weeks. One quarter were applied shortly after receiving a shipment, and a second quarter was applied a week later. The remaining 25,000 mites were kept in their container but turned on the side until they were applied during the third or fourth week. The mites hold up pretty well if stored at 65° to 70°F days. It was difficult to hold the mites longer than two weeks because the environment in the container became unsuitable and mites began migrating out. The cost for 50,000, including shipping, was $38.25. Total cost of mites from March through May (four shipments) was $153.00.

We applied the BotaniGard using a Dramm cold fogger at 3,000 psi. In the greenhouse where a combination of mites and BotaniGard was used, the mites were applied first, then we applied BotaniGard.

How Did It Work?

In the early part of the study, thrips activity was low on the sticky cards. On April 20, we started to see significant differences on the number of thrips on sticky cards in control greenhouses compared to the treatment greenhouses. Sticky card counts remained low throughout the study in the greenhouses receiving predacious mite releases, combinations of BotaniGard and predacious mites, and BotaniGard alone.

In the three control houses, the counts of thrips was relatively low for the first two turns of bedding plants, but by the third crop turn the thrips counts averaged .75 thrips, 1.0 thrips, 1.25 thrips per plant, respectively.

The greenhouse receiving *Beauveria bassiana* alone remained very low through the first two plant turns of the study, with under .16 thrips per plant but rose slightly at the end to .5 thrips per plant. The predacious mite–release greenhouse kept at zero for the first two plant cycles with a slight rise to .15 thrips per plant during one week. This level dropped to zero by the time the plants were moved to market. The greenhouse receiving the combination of predacious mites and *B. bassiana* was very similar to the greenhouse receiving just the predacious mites.

Based on this demonstration, we feel confident that an early release of predaceous mites, or a combination of mites and *B. bassiana*, evenly distributed through a bedding plant crop, can provide a good way for growers to effectively manage thrips populations in early spring bedding plant production.

Pest Management, February 2000.

Biological Thrips Control on Pot Sunflowers

Stanton Gill

Sunflower plants are magnets for thrips when grown in greenhouses. We decided that trying to keep thrips at below damaging levels on sunflowers would be a good challenge. We used *Helianthus annus* 'Big Smiles' in trials at Bell Nursery, Burtonsville, Maryland. A dwarf sunflower, 'Big Smiles' grows eighteen to twenty-four inches tall as a potted plant.

We planted three sunflower seeds in eight-inch pots on June 15, 1996. Two greenhouse bay sections separated by a partition wall were used in the field trial. There was no microscreening on the intake vents, so outside and inside pressure from thrips was very strong. We identified two predominate thrips species, western flower thrips, *Frankliniella occidentalis*, and flower thrips, *Frankliniella tritici*, on the crops.

Because this is a working greenhouse, we couldn't leave untreated plants for controls. In this trial, we compared *Beauveria bassiana* applications to standard chemical treatments for suppressing thrips on potted sunflowers. Greenhouses treated with *B. bassiana* (BotaniGard) received five applications. A second greenhouse separated by a greenhouse bay and wall partition received chemical treatments to control thrips. We let the grower select chemicals to use for thrips.

In each greenhouse, ten plants were randomly selected from a 2,000-sq. ft. growing area. Each week an IPM scout used a clipboard with a white paper and performed a tap test on plant foliage. All larval and adult thrips counts were recorded. When plants came into flower, flowers were shaken vigorously three times, and the numbers of dislodged live and dead thrips were recorded. Thrips removed from blooms were collected and taken back to the Central Maryland Research and Education Center lab, where they were dipped in alcohol and placed on a petri dish with prepared agar. We held the thrips at 72°F for seven days and recorded the number infected with *B. bassiana*. Infection was determined by fungus growth on the body of the thrips.

B. bassiana was applied at 2 lbs. per acre. We mixed 40 g BotaniGard with 6 ml SilWet and shook the container. The solution was then mixed with 12 gal. of water. This would cover a 2,000-sq. ft. bench area. The spray was applied in a fine mist to coat upper and lower leaf surfaces. We applied BotaniGard on July 2, July 9, July 17, July 20, and July 22, 1996. On July 29, 1996, thrips numbers on foliage and flowers were still at an objectionable level for sales, so we applied Dycarb to all plants to bring thrips down.

The greenhouse receiving chemical treatments for thrips control required three applications. Applications of Avid/Mavrik were made on July 7 and July 29. A third chemical application was made on July 22 using Dycarb.

Results

In cases where pressure from existing and inward migrating thrips populations is strong, both chemical controls and *B. bassiana* brought thrips populations down but didn't totally suppress them. Once flowers had opened, thrips counts on plants increased substantially in both greenhouse treatment bays. *B. bassiana*–treated plants and one Dycarb application gave about equal control to the three chemical treatment greenhouses.

Thrips in Chemical-Treated Greenhouse

Plant number	Live thrips counts					
	July 1	July 8	July 16	July 24	July 30*	August 5*
Range for 10 pots	2-18	2-25	4-15	1-11	2-25	1-13
Total for 10 pots	86	92	83	38	74	64
Average for 10 pots	8.6	9.2	8.3	3.8	7.4	6.4

* By 7/30 plants had flowers open, and most thrips were found in flowers.

Thrips in *Beauveria bassiana*-treated greenhouse

Plant number	Live thrips counts					
	July 1	July 8	July 16	July 24	July 30*	August 5*
Range for 10 pots	3-11	2-13	4-15	0-6	1-13	2-24
Total for 10 pots	56	71	63	26	51	92
Average for 10 pots	5.6	7.1	6.3	2.6	5.1	9.2

*By 7/30, plants had flowers open, and most thrips were found in flowers. When these thrips, recovered on 8/5, were placed on petri dishes, 29% showed infection with *B. bassiana.*

Thrips tapped from *B. bassiana* plants were dipped in alcohol and placed on agar in petri dishes to detect how many were infected with the fungus, as judged by hypal growth of the fungus from the body cavity of the thrips. Twenty-nine percent of thrips recovered were infected with the fungus. It's possible that other uninfected thrips weren't contacted by the spray because of misdirected application technique or that these thrips migrated into flowers after the last application was made. The grower judged all plants in the *B. bassiana*–treated bay and the chemical-treated bay salable and sold them by August 6, 1996.

Culture Notes, August 1997.

Whiteflies

Whitefly Battles: Expand Your Options

With the rising danger of pests developing resistance to favorite chemicals used for pest control on poinsettias, biological control becomes an even more important tool for management, particularly for poinsettias' common foe, whiteflies. Many options are available, and poinsettias are an excellent crop for biological control. Natural enemies for poinsettia pests are compatible with commonly used fungicides and growth regulators, and effective chemicals are available for backup if needed, according to Christine Casey, University of California, Davis, in Cornell University's "Greenhouse IPM Update." Here's an overview of your biological control options for whiteflies.

Encarsia formosa

Encarsia formosa, a parasitic wasp, attacks greenhouse whitefly and silverleaf whitefly, most effectively controlling greenhouse whitefly. You'll receive the wasp as parasitized whitefly pupae glued to cards. Place the cards face down as close to the center of the bottom of the plant as possible. Wasps fly in an upward spiral as they emerge, so they'll find nymphs on leaf undersides as they fly upward. They're most effective at 80°F and 50 to 80% relative humidity. Pesticides aren't compatible, but soap, oil, fungicides, and insect growth regulators are fine. Consult your supplier for recommended release rates.

Eretmocerus eremicus

Eretmocerus eremicus, also a parasitic wasp, attacks silverleaf whitefly, though it will also attack greenhouse whitefly. You'll receive batches of one thousand parasitized pupae. Divide them into groups of fifty; put them in release cups and put them uniformly throughout the greenhouse as close to the center of the plant as possible. Suggested release rate is three to five wasps per plant per week, though your supplier may suggest a different rate for plant size and density. Like *Encarsia,* they can't survive pesticides but are unaffected by soap, oil, insect growth regulators, and insect fungicides. At 77° to 84°F, the wasp will develop faster than silverleaf whitefly. Note: *Eretmocerus* likes yellow sticky cards, so reduce the number of cards you put out when using this wasp.

Delphastus pusillus

The predatory beetle *Delphastus pusillus* is most successful in high whitefly populations as a "hot spot" control used with other biological controls. They will reproduce if they consume large numbers of whitefly eggs. Don't release them near yellow sticky cards, as they're attracted to them. Release rates vary with different whitefly levels, plant sizes, and spacings.

Beauveria bassiana

The only fungus in our list, *Beauveria bassiana*, is sold under the biopesticide names Naturalis-O and BotaniGard. Healthy whiteflies pick up fungal spores as they walk across infected insects, spreading the pathogen. Infected insects can turn orange-brown. *Beauveria* also has the unique ability to infect at relative humidities down to 45%. It isn't compatible with fungicides.

Culture Notes, October 1997.

Weapons against Whitefly

John P. Sanderson

Whiteflies are pests mainly because consumers don't want to see them or any other insect on the plants they buy. In high whitefly populations, their secretions of honeydew and the ugly gray sooty mold that grows on it can destroy a plant's aesthetics. A severe infestation of whitefly can cause poinsettia bracts and stems to turn whitish yellow. Whitefly can transmit geminiviruses on outdoor vegetables and can transmit tomato yellow leaf curl virus in greenhouse tomato transplants.

Now that your greenhouses are full of poinsettias, it's time for a refresher in whitefly control.

Identification and Biology

The silverleaf whitefly (SLWF), *Bemisia argentifolii*, and the greenhouse whitefly (GHWF), *Trialeurodes vaporariorum*, are the most common whitefly pests.

Females can lay two hundred or more eggs and live up to one and a half months. All life stages are found on lower surfaces of leaves. Tiny scalelike crawlers walk a few millimeters from the egg, insert mouthparts into the leaf to feed, and don't move again until they've completed the remaining three nymphal life stages and emerge as adults. On poinsettias at 65° to 75°F, total egg-to-adult development takes thirty-two to thirty-nine days. Development time is considerably faster at warmer temperatures, perhaps by two and a half to three weeks. Eggs are immune to most insecticides; fourth-instar pupae can also be difficult to kill.

Monitoring

The weekly use of yellow sticky cards (one card per 1,000 sq. ft.) for adults, coupled with plant inspections for nymphs, can give a good overall picture of the presence, size, and location of an infestation and reveal if control strategies are working.

Management

Season-long control will be much easier if you start clean. Find and eliminate existing infestations (weeds, unsold hanging baskets, old stock plants, "pet" plants, outdoor gardens near your greenhouse). Maintain thorough weed control beneath benches and outside the greenhouse. In warmer climates, screened doors and vents can reduce migration into the greenhouse. But remember that whiteflies can still enter a screened greenhouse on incoming plant material, on workers' clothing, and through doors that are left open.

Chemical Control

Marathon (granular or drench) continues to give excellent long-term control when used properly. Apply the product after you have good root growth (roots that have grown to the edge of a pot), and don't allow water to leach from the bottom of the pot, washing the insecticide out of the container, for a week after application.

Many other insecticides are registered for whitefly control. Many growers find foliar, aerosol, or smoke applications helpful before Marathon is applied or near the end of the crop if a cleanup is needed (apply according to the product label to avoid plant damage—some can damage poinsettia bracts). Mixtures of a pyrethroid such as Tame with Orthene are still effective for many growers.

At least two new and very effective insect growth regulator (IGR) insecticides are now or soon will be available: buprofezin (Applaud by AgrEvo) and pyriproxifen (Distance by Valent or PyriGro by Whitmire Micro-Gen). Both are for foliar application. Pyriproxyfen has translaminar properties (it moves through the leaf surface), long residual control, and is especially effective on young nymphs and eggs. Buprofezin has substantial vapor activity that aids in "coverage." Both show excellent activity against nymphs (not adults) and are very compatible with whitefly parasitoids. These products must be used carefully, according to label directions, or resistance problems are likely to occur. They provide important new insecticide options for pesticide rotation schemes.

Biological Control

Whitefly biological control could include the release of parasitoids and/or predators and/or fungal pathogens. For biological control to be successful, rely on releases of the natural enemies and use selected insecticides as a backup. Growers interested in biological control would need to have established a successful whitefly monitoring plan.

Encarsia formosa is the most commonly used parasitic wasp for GHWF on greenhouse tomatoes. *Eretmocerus eremicus* is another parasitoid that provides better SLWF control on poinsettia and also will control GHWF. For successful SLWF management with parasitoids alone in the temperate regions of the U.S., *E. eremicus* should be released weekly at three female wasps per pot per week. But such a release regime is quite expensive. A less expensive approach may be to release *E. eremicus* at one female wasp per pot per week, coupled with an IGR (pyriproxyfen or buprofezin) applied once, just before bracts begin to color.

BotaniGard and Naturalis-O contain spores of the insect fungal pathogen *Beauveria bassiana*. This pathogen should be used while whitefly levels are still very low. Three to five weekly sprays should be applied with a hydraulic sprayer, targeting the lower surfaces of leaves—then carefully evaluate the control to determine the need for additional sprays. Tank mixes with most conventional insecticides can be used to reduce pest levels, but do not mix with any fungicides and be sure that the spray tank is clean of all fungicide residues. Do not use forty-eight hours before or after a fungicide application on the crop.

Pest Control, September 1998.

New Chemistry Combats Whiteflies and Mites

The whitefly is the most economically damaging pest across all segments of ornamental production. It has been cited as the No. 1 pest by 36% of growers in a survey of 300 ornamental producers. Mites were cited by 16%.

So it's understandable that growers have a very low tolerance for these pests in their greenhouses. But with only a handful of products registered for greenhouse ornamentals, overuse can occur, causing pests to become resistant. However, a new miticide/insecticide from BASF Agricultural Products, Research Triangle Park, North Carolina, lets you rotate an entirely new chemistry into your insect control program.

Sanmite miticide/insecticide is a broad-spectrum miticide with effective residual control of several mites and excellent control of whiteflies. It's labeled for control of two-spotted spider mites, broad and European red mites, as well as silverleaf and greenhouse whiteflies on twenty-six greenhouse ornamentals and flower and foliage crops, including poinsettias, chrysanthemums, and roses.

In trials conducted in Florida in 1995 on various ornamental species, overall mortality of two-spotted spider mites was 95% two weeks after treatment. Mortality of adults averaged 83%, but represented only 10% of the pest population. Mortality of larvae and nymphs was 95%.

Although Sanmite will not harm even sensitive ornamentals, not all plant species have been tested with possible tank mix combinations. Before using Sanmite, growers should test a sample of the crop to ensure no phytotoxicity will occur.

How It Works

Sanmite contains the active ingredient pyridaben, developed by Nissan Chemical Industries Ltd. Pyridaben is a contact miticide/insecticide and, as such, requires thorough coverage of the underside of leaves to ensure pests will come in direct contact with the product. This unique chemistry works as a mitochondrial electron transport inhibitor (METI) to block cellular respiration, causing the pests to lose motile coordination and eventually collapse.

"This is not only a new mode of action, but a new site of action because it's not working on the nerve, as the traditional miticides do," said Vivienne Harris, BASF field biologist.

Resistance Management

This new chemistry is good news because growers are facing resistance with current chemistry. According to a July 1994 market study for BASF, 42% of growers surveyed suspected whitefly resistance in their greenhouses and 18% suspected mite resistance to one or more products. This is one reason BASF has introduced Sanmite with a plan for resistance management.

"The advantage of a totally new chemistry, such as pyridaben, is its high effectiveness and long residual control," Vivienne said. "To extend this advantage, BASF has labeled Sanmite for use only in a rotation with at least two other products between applications. While adding Sanmite to their insecticide rotation, growers also need to review their current program,"

Vivienne adds. "Growers need to make sure they are rotating modes of action, not just brand names."

To reduce resistance, growers need to rotate chemical classes, such as chlorinated hydrocarbons (Pentac, Kelthane), pyrethroids (Mavrik, Talstar), nicotinic receptor blockers (Marathon) and avermectin (Avid), as well as a METI such as Sanmite.

Using Sanmite in a rotation will help prevent resistance, as well as break the cycle of resistance to other products by using them less frequently. To further reduce resistance, growers should spray insecticides only when needed and also spray spots where infestations are more severe, rather than spraying all plants.

Mite/Whitefly Life Stages

Although all stages of the mite's life cycle are affected by Sanmite, certain stages are more susceptible than others. For example, two-spotted spider mite larvae and nymphs are more susceptible to pyridaben than adult females. Adult females are usually the only lifestage noticed by most growers because they are the biggest and move around the most.

Adult whiteflies and nymphs and adults are controlled by Sanmite better than the pupae and eggs are. Because whitefly adult females live for ten to thirty days and oviposit throughout their life span, populations on ornamentals usually are a mixture of all life stages. This makes the timing of a Sanmite spray less important. Although some whitefly pupae, the least susceptible stage, will survive, once they molt and become adult, they will die. Adult whiteflies are very susceptible, and they soon pick up a toxic dose.

"If you have a severe adult infestation of mites or whiteflies, it would be best to use an inexpensive, smothering-action product, like insecticidal soap, for immediate knockdown, then follow with an application of Sanmite for residual control," Vivienne said. "However, it's best to target mites and whiteflies when the populations are building and not wait until the populations are dangerously large."

Residual Control

BASF recommends using Sanmite at 2 to 4 oz. per 100 gal. of water to control mites and 4 to 6 oz. per 100 gal. of water for control of whiteflies. Once it has dried on the plants, Sanmite has the ability to persist on leaf surfaces for several weeks, despite overhead irrigation or rain. When sprayed at labeled rates with good plant coverage, Sanmite consistently controls mites for three weeks or more at the 4-oz. rate and whiteflies for five weeks or more at the 6-oz. rate.

With such long-lasting control, evaluating its effectiveness at a week or less after treatment does not give an accurate picture of Sanmite's control. As with any contact miticide/insecticide, it's almost impossible to kill an entire mite or whitefly population immediately upon spraying. Time is needed for the mobile individuals not hit directly with the spray to move around and pick up a toxic dose from the plant surface.

Many times growers check their plants the next day expecting to see a drastic reduction in the population. That won't happen with Sanmite. The number of immature mites will be cut by more than 90% within a week. But some adult females will still be noticeable, especially where the pretreatment population is large. Check again about ten days after spraying, and they'll be gone, too. Meanwhile, Sanmite's residual activity will control any larvae hatching from eggs.

Continue monitoring the pest population after applying Sanmite, and when the mite/whitefly population begins to reemerge, apply a different product. When the population rebuilds after the second insecticide application, the third chemical in the rotation should be applied, and so on. Remember to make sure the insecticides all use different modes of action to ensure the best control and to forestall resistance development.

Editor's note: Information provided by Fleishman-Hillard Inc. for BASF Agricultural Products, Research Triangle Park, North Carolina.

January 1997.

Wildlife

Managing Wildlife Pests in and around Greenhouses

Paul D. Curtis

Several wildlife species may cause problems in and around greenhouses. To reduce damage, you should know the habits and environmental factors influencing animal pests and use the proper management methods.

Exclusion

Mice or voles often invade greenhouses when the weather turns colder during October and November. In the fall, rodent densities are near their annual peak, and food and/or cover provided by annual vegetation becomes scarce after a few frosts. Rodents begin moving from their spring and summer nesting sites into buildings in search of food and shelter.

Exclusion is the best long-term solution for managing troublesome rodents. Mice or voles will enter greenhouses through openings as narrow as a quarter-inch (rats require about a half-inch). Mice are also good climbers, and all points of entry should be sealed to about three feet above ground. The most common access point for rodents is the door. Easy-to-install door sweeps can be attached to bases of doors with screws. Also, never prop doors open.

Building maintenance is very important for successfully excluding rodents. The points where pipes or wires enter buildings or penetrate walls are other potential entry areas for mice and rats. Galvanized metal pipe chase covers can be used to seal gaps around pipes in walls. Also, repair greenhouse foundation holes and broken floor drains, screens, and windows. Seal small openings and narrow gaps with inexpensive latex caulk applied with a reusable caulk-gun. Steel wool or flexible, aluminum gutter guards can be wedged into gaps and holes using a screwdriver. Many inexpensive materials can be used to rodent-proof structures.

Managing the habitat outside of the greenhouse can also reduce rodent numbers and potential damage. Waste lumber and piles of scrap materials around greenhouses can provide a haven for rodents, which move into structures when food supplies and weather change. Remove thick vegetation (weeds, bushes, vines), debris (litter, garbage, leaves), and clutter (bricks, rock piles, pallets) that provide cover. A rule of thumb is to maintain at least a one-foot vegetation-free border next to the greenhouse using gravel as a barrier to reduce weeds and prevent rodents from digging under the foundation.

Trapping

Sometimes it's necessary to directly reduce animal numbers within a greenhouse. Trapping is a time-tested method for controlling rodents. Snap traps and glue boards are most commonly used. Place snap traps along established travel routes, with the trigger facing toward the wall. Traps may also be attached along pipes and rafters using wire or Velcro strips. Place peanut butter (for mice) or apple (for voles) on the center of the trigger or glue board. Traps placed in areas with overhead cover (such as boxes or coffee cans) usually have higher capture rates. If rats are the problem, it may be necessary to pre-bait (provide baited traps without setting them) for a few nights because rats have a fear of new things. Multiple-catch mousetraps, available in hardware stores, can quickly reduce rodent numbers.

Successful trapping involves an understanding of animal habits and characteristics. Rodents are most active after dark and before dawn, and their presence is usually revealed by feces or gnawing activity. Mice deposit forty to sixty fecal droppings per day, so a large number of droppings doesn't always indicate many mice. Rodents may gnaw on wood, plastic irrigation lines, or wire sheathing, and their presence is announced when growers find the shredded remains of these materials. Mice gnaw holes approximately one inch in diameter, and rat burrows measure two inches or more. Voles typically form runways under pallets or flats that provide overhead cover.

Chemical Controls

Rodenticides are classified in two major groups based on whether they're single dose and fast acting (acute), or multiple dose and slower acting (chronic or anticoagulant). Regardless of which type of bait you use, always use tamper-resistant bait stations when children, pets, domestic animals, or nontarget wildlife are at risk. It's generally much safer and more effective to use anticoagulants rather than acute toxicants for rodents.

The effects of anticoagulants are cumulative and most must be consumed over a number of days to produce death—that's why they are considered less hazardous for humans and other nontarget animals. Rodents also readily accept them; rarely do rats or mice develop a shyness to good-quality anticoagulants.

Anticoagulants can be used in bait blocks or throw packs (for inaccessible areas). For the greatest effect, rodents should feed on the treated bait each day, and the intervals between feedings shouldn't exceed two days. Depending on the anticoagulant and the amount consumed, rodent deaths should occur about the fourth day after first exposure. Provide fresh rodenticide continuously for at least two weeks or until all feeding ceases. With some anticoagulants, shorter periods of exposure are recommended, so be certain to follow all label instructions for specific products. In some states, these rodenticides are registered for voles as well as rats and mice.

Repellents

A variety of commercial repellents are available to reduce deer or rabbit browsing. The effectiveness of repellents is extremely variable and is affected by factors such as deer or rabbit numbers, feeding habits, and environmental conditions. Repellents may be cost effective for controlling wildlife damage when light to moderate damage is evident, small acreages are damaged, and three or fewer applications are needed for adequate control. If these three conditions aren't satisfied, a grower may want to look at the cost-benefit ratios of electric fence designs or other alternatives.

With the use of repellents, some damage must be tolerated, even if browsing pressure is light. Apply repellents before damage is likely to occur and before a feeding pattern is established. Make repeat applications of repellents every four to five weeks while plants are susceptible to damage. Apply repellents when precipitation isn't expected for twenty-four hours and temperatures will stay at 40° to 80°F. Applications should be thorough, covering all plant parts.

If browsing pressure is severe, growers should evaluate exclusion alternatives. Fencing is the most reliable technique for preventing damage from medium to large mammals. Woven-wire designs are the most effective physical barrier to wildlife, with high-tensile, woven-wire fencing providing the ultimate in protection and durability. Also, a variety of multi-strand, vertical, or sloped electric fence designs can effectively exclude wildlife. Electric fences may be complete physical barriers or, more commonly, may act as psychological deterrents. Electric fences won't exclude wildlife unless adequate voltage is constantly maintained on the wires. Disadvantages include frequent monitoring and the need for vegetation control to maintain shocking power.

Greenhouse operators have many options for dealing with problem wildlife. Consider the level of damage that's tolerable, then determine the cost effectiveness of methods that will produce the desired results. There is no simple solution that will solve all wildlife conflicts, and damage management programs must be adapted to each specific situation.

Pest Control, October 1997.

Chapter 7
General Disease Control

Disease Control Tips from the Experts

Dr. Roger C. Styer

At the Society of American Florists' Thirteenth Conference on Insect and Disease Management in Orlando, Florida, some of the country's foremost disease researchers offered control updates and tips you can put to work in your greenhouse.

Margery Daughtrey, Cornell University, Long Island Horticultural Research Lab, Riverhead, New York, reviewed four key diseases that she has seen more of in the last few years. They include *Cercospora* leaf spot on pansy, *Alternaria* leaf spot on impatiens, *Phytophthora* on a number of crops, and, of course, impatiens necrotic spot virus (INSV) on many crops. In addition, Dr. Gary Simone, University of Florida, Gainesville, covered *Acidovorax* leaf spot, a new disease on geraniums.

Cercospora Leaf Spot on Pansy

This disease is caused by *Cercospora violae*, a fungus specific to pansy. It can be confused with other fungal leaf spots that affect pansy, so proper lab diagnosis is important. Symptoms include large, spreading, purple lesions on lower leaves, which dry and turn necrotic. Spots are irregular in both size and shape and don't have a sharp, rounded outline. This disease may be seed-borne, but more research is needed to confirm this. It has shown up in plug production during warm summer and early fall months and has caused significant losses after shipping and transplanting infected plugs. Recommended fungicides include Cleary's 3336, FungoFlo, Daconil, Protect T/O, and Zyban.

Alternaria Leaf Spot on Impatiens

Symptoms are similar to those of INSV and *Pseudomonas* leaf spot, so get a lab diagnosis. *Alternaria* spots are tiny, purple specks when they begin to develop; they become dark-rimmed with a pale center when mature. The leaf blade often becomes chlorotic near the lesions. This is different from INSV or *Pseudomonas* leaf spot. Lesions are uniformly round, rarely more than an eighth-inch in diameter and are scattered across the leaf rather than being concentrated on margins, as is the case with *Pseudomonas* leaf spot. Control *Alternaria* by roguing out infected plants and controlling leaf wetness. Recommended fungicides include Chipco 26019, Daconil, Phyton 27, Kocide, Dithane, Manzate, and Protect T/O. Avoid using chemicals such as Cleary's 3336, FungoFlo, and Zyban, as thiophanate-methyl materials may exacerbate the problem.

Phytophthora parasitica
This disease has a wide host range and can spread in water sources, by splashing water and on insects. It can also persist in the landscape. *Phytophthora* can be misidentified if growers are trying to control *Rhizoctonia, Botrytis, Thielaviopsis,* and leaf spots. Problems have shown up on pansy and vinca plugs and on fuchsia and poinsettia cuttings. Scout for and remove diseased plants. Minimize spread by shore flies and fungus gnats, and reduce the survival of inoculum in greenhouses. Recommended fungicides include Subdue, Aliette, Terrazole, Truban, and Banrot.

INSV
This virus is the most serious problem greenhouse growers encounter. It has a wide host range, produces different symptoms on different plants, can produce no symptoms but still infect plants (asymptomatic) and is spread by western flower thrips. INSV can easily be spread from greenhouse to greenhouse through propagated material such as cuttings or prefinished plants and can persist in crops or weeds in the greenhouse. Using on-site test kits (QTA-Tospo kits from Agdia) can help you determine the presence of the virus. Use indicator plants and sticky cards to quantify thrips populations and whether they're carrying the virus. Be aggressive with your thrips control program. Control weeds in and around the greenhouse. Get rid of any infected crops immediately, and don't hold leftover crops when the season is finished. Clear out vegetative stock every year, and start with new material.

Acidovorax Leaf Spot on Geraniums
This new bacterial leaf spot on seed and vegetative geraniums has shown up in Florida, Michigan, and Indiana. It's difficult to diagnose, as *Botrytis* and other bacteria may be associated with it. *Acidovorax* is especially associated with young plants (plugs and cuttings) and can be found on petunia, impatiens,

vinca, verbena, and foliage plants. Gary Simone has initiated a research project to find out more about this new bacterial leaf spot.

Culture Notes, June 1997.

Update on Disease-Suppressive Potting Mixes

Raymond A. Cloyd and Harry A. J. Hoitink

In the early 1970s, most potting mixes were prepared with soil or sand, peat, and perlite, often at 1:1:1 ratios. These types of mixes had to be steamed to kill weed seeds and plant pathogens. After potting, crops highly susceptible to root rot, such as poinsettias and Easter lilies, had to be drenched with fungicides on a regular schedule to prevent soilborne diseases such as *Pythium, Phytophthora, Fusarium, Rhizoctonia,* and *Thielaviopsis.*

Today, several types of potting mixes have been introduced that naturally suppress some or all diseases caused by these pathogens. The first of these disease-suppressive potting mixes that was widely distributed contained aged or composted tree barks. Hardwood bark mixes were most effective for disease suppression, but they supported too much microbial activity. As a result, fungus gnat populations were stimulated because they thrive on the naturally introduced microflora.

Composted pine bark and fir bark mixes have become more widely used by greenhouse growers. The lignin (brown material) and protected cellulose (white-yellow material) in composted bark mixes can support the beneficial microorganisms throughout a long-term poinsettia or cyclamen crop. The microorganisms in these mixes bring about these beneficial effects through competition with pathogens, antibiotic production, parasitism or predation (literally by eating plant pathogens), or by inducing plants' systemic resistance to diseases. Some biocontrol agents, such as specific strains of *Trichoderma,* can express all four mechanisms.

During the 1980s, a major breakthrough was made in the harvesting and formulation processes used for sphagnum peat mixes. Several laboratories discovered that light sphagnum peat harvested from the top two to four feet of peat bogs became naturally suppressive to *Pythium* root rots shortly after potting-mix formulation. The darker sphagnum peat, which is more decomposed and comes from deeper in peat bogs, can't support this beneficial effect, even when inoculated with biocontrol agents. This has been shown many times by plant pathologists. Thus, fungicide drenches are the only way to control many types of root rots in dark peat mixes.

Another important development in the peat industry was the introduction of star-screening systems. Star screens allow the peat industry to remove fine particles from peat and prepare products more consistent in particle size. As a result, the peat industry can now prepare potting mixes with ideal drainage and water-retention properties that also provide some degree of natural biological control of diseases.

The new sphagnum peat technology and bark mixes have allowed growers to apply fewer fungicide drenches. Poinsettias may have to be drenched only once, and many crops aren't drenched at all. These natural mixes have been more effective on *Pythium* and *Phytophthora* root rots. The effect has been better for control of *Fusarium* wilt on cyclamen than fungicides achieve. However, for *Rhizoctonia,* fungicides still are more reliable.

Several biocontrol agents, such as *Trichoderma* and *Streptomyces* preparations, that can improve the efficacy of natural mixes, are now available.

For composted pine bark, the story is the same. Some bark mix producers "compost" their bark in large piles at excessively high temperatures. When temperatures during composting range from 100° to 140°F, an excellent-quality product can be produced. In large piles, however, temperatures may reach 160°F or even higher. Under these conditions, the bark burns up or turns black (pyrolysis), losing its lignin and protected cellulose and no longer supporting the activity of biocontrol agents. These black bark particles also may lose their physical integrity and change to powder (charcoal) during the mixing or potting processes. This destroys the physical properties related to drainage, and *Pythium* root rot is sure to become a problem on susceptible crops.

As you can see, the idea of using potting mix to control root diseases has made significant progress during the past two decades. But there's still much to be learned, and growers need to understand how these soils work and what pitfalls may lurk. For more details, read "How to Optimize Disease Control by Composts" by Hoitink, Jaber, and Krause in *Tips on Managing Floricultural Crop Problems*, available through the Ohio Florists' Association.

Pest Management, March 2000.

Soil Treatment for Disease Control

R. Kenneth Horst and Jim Locke

Common greenhouse crop pathogens range from relatively easy to control to more difficult to control. Many root pathogens fall into this latter category, especially where beds or benches are replanted with the same crop. These pathogens, which cause damping-off, root rot, and wilt, survive in growing media and require treatment to eliminate them or at least greatly reduce their numbers. Steaming is just one option available to treat soil and growing media to reduce unwanted pests and pathogens.

Steaming Bulk Soil

Many pot plants and bedding crops are grown in commercially produced soilless mixes. It is generally recommended that these mixes not be steam sterilized because of detrimental effects on microbes in the mix. However, when most growers used field soils in greenhouse production, sterilizing soil (usually with steam) was a requirement.

Steam sterilizing greenhouse soils has been considered the best and most effective method of eliminating soilborne pathogens. In addition, steam kills weeds and other pests. Other advantages of steam sterilization include no phytotoxicity beyond the immediate treatment area and no harmful effects to plants in the same greenhouse. Little aeration time is required; steamed soil can be planted as soon as it's cool. Also, it's adaptable to many situations.

There are various methods for steaming soil. Boilers that heat greenhouses can be adapted to supply steam for sterilizing benches or soil bins, but you must consider boiler-heating efficiency. A new boiler could have an 80% efficiency rating, whereas a twenty-five-year-old boiler may have only a 60% rating. The heat-exchange efficiency of a particular soil can have a significant effect on the number of BTUs required for steam sterilization. Heat exchange efficiency is sometimes difficult to determine. Properties that affect heat movement in soils include soil type, texture, and moisture content. As a conservative estimate, 2,500 BTUs would be required to raise a cubic foot (0.028 m^3) of loose, moist soil with a heat-exchange efficiency of 60% from 62°F to 212°F.

Another option is a commercially produced soil-steaming wagon. Bouldin & Lawson builds these on a custom basis. Portable steam generators are also available. The Steam-Flo Generator from Sioux Steam Cleaner Corp. is fully equipped with oil burner, complete controls, and safety devices. It's on wheels and can quickly be moved into place. Steam is available in twenty to twenty-five minutes from a cold start. It will sterilize a five-by-fifty-foot bench in two hours, maintaining eight to ten pounds pressure while burning four to four and a half gallons of No. 1 fuel oil per hour with a 425,000-BTU output. The Mayer IMO Soil Steamer System from Gro-May Corp. is suitable for mobile and stationary use and has standard dual steam (up to 392°F at an operational pressure of one bar) or warm-water heating, providing an economic and environmentally friendly alternative to soil fumigation. With the ban on methyl bromide use in the U.S. in the year 2001 under the Clean Air Act, we may see increased use of steam generators.

The following table will help you calculate and compare the costs of steam sterilization and other methods of soil treatment. Available BTUs per unit of fuel is a more useful common point of comparison than price per unit, as price changes over time and with geographic location. Formulas for cost determination include soil volume, soil heat exchange efficiency, boiler efficiency, units of fuel required, and the BTU constants of the fuel. Knowing the price of the fuel, you can then calculate the cost per measure of soil.

Equations for Calculating Fuel Required to Sterilize Soil

V = volume of soil in cubic feet = bench length x width x depth
E = energy in BTUs required to raise soil temperature from 62° to 212°F
   E per ft^3 loose, moist soil = 2,500 BTUs
S = soil heat exchange efficiency
B = boiler efficiency
U = BTUs per unit of fuel
   U No. 2 fuel oil = 140,000 BTUs per gal
   U natural gas = 100,000 BTUs per therm
   U coal (anthracite) = 13,000 BTUs per lb.
F = units of fuel needed = $(V \times E) / (S \times B \times U)$

Sample problem: Calculate the amount of fuel needed to sterilize a greenhouse bench with 200ft^3 of soil using No. 2 fuel oil and a boiler of 80% efficiency.
   F = (200 ft^3 x 2,500 BTUs) / (0.6 x 0.8 x 140,000 BTUs per gal.)
   = 500,000 BTUs / 67,200 BTUs per gal.
   = 7.44 gal. fuel oil

Fungicides from Nature

Chris Beytes

A new soilborne disease control product has hit the market recently, and at least one major grower is calling it phenomenal.

BioWorks' RootShield biological fungicide uses a hybrid strain of fungus called T-22 to prevent root diseases. According to Dave Pieczarka, BioWorks' director of sales and marketing, RootShield's history started twenty-two years ago when a Cornell University researcher began identifying naturally occurring fungi that protect plant root systems from diseases. He found hundreds of such fungi in soils around the world and began a breeding program to develop vigorous hybrids. The result is strain T-22, which is one hundred times more effective at protecting plants from disease than is the naturally occurring strain. Cornell patented it, and BioWorks has the license to market the product worldwide.

RootShield is available as a drench for top application or as a granular for growers who mix their own soils. Because it's a living entity, the fungus needs temperatures above 50°F and adequate moisture to survive and reproduce. RootShield lives on plant roots, existing off of what the roots exude. In return, RootShield protects plants from soilborne diseases including *Pythium, Rhizoctonia, Fusarium,* and *Sclerotinia.*

How well does it work? "Phenomenally," says Bill Swanekamp, owner of Kube-Pak, Allentown, New Jersey. Since first trialing the product last October, Bill has begun adding it to every crop he grows.

Bill first tested RootShield on unrooted fuchsia cuttings, a tough crop to propagate, especially in the fall. Normally expecting a 5 to 10% loss, he lost almost none of the RootShield-protected cuttings. He then tried it on unrooted New Guinea impatiens, vinca vine, and begonia cuttings, again with excellent results. Bill was so sold on the product that he used it on his entire production of 120 million plugs, along with 350,000 flats of bedding plants, all of his hardy mum cuttings and finished pots, all of his direct-stuck poinsettia cuttings, and twelve acres of finished poinsettias. He says he's almost eliminated all fungicide drenches.

The costs of incorporating RootShield into the soil are comparable to the costs for chemicals and labor for fungicide drenches, Bill says. But there's no reentry period, and you don't have to have employees working after hours to apply the product. Payback comes from the increased production he's been getting—for instance, he used to direct stick 10 to 15% more ten-inch poinsettias (four cuttings per pot) than he needed to make up for losses in propagation. With RootShield, "This year I've got more ten-inch poinsettias than I know what to do with," he says.

Bill says there's an added benefit: more vigorous growth. He's actually shortened his production time by a week on crops such as *Celosia*. He's cut losses of unrooted cuttings to almost nothing. And his plug customers say this year's plugs were the best they've ever received from Kube-Pak. Needless to say, Bill's sold on the product.

One caution: RootShield is a protectant, not a curative. Bill stresses that you've still got to watch for disease problems and maintain good water management and sanitation practices. Also, RootShield won't protect against foliar diseases such as *Botrytis*.

RootShield's granular form takes effect immediately; the drench form must be applied at sow or transplant. It lasts three to four months. Long-term crops such as perennials need a booster application every six months. Despite being a fungus, RootShield is compatible with most commercial fungicides—Dave says it's very tough to kill.

GrowerTalks in Brief, December 1997.

Perennial Disease Watch

Cheryl Smith

Now is the time your perennials are at risk for disease infestation, but knowing what to watch for is half the battle. Here we've put together a quick guide to common diseases. To prevent disease resistance when using fungicides, rotate products from different chemical families.

Botrytis Blight

Common hosts: Aster, rudbeckia, peony, and salvia.

Monitoring: Wounded, older, and young succulent tissues are particularly susceptible. May cause leaf spots, leaf blights, stem cankers, cutting rots. Under high humidity, a fuzzy gray, gray-green or gray-brown "mold" may develop.

Management: Sanitize; remove and destroy infected tissues. Increase plant spacing for air circulation; remove older plant parts; deadhead plants; remove all crop debris; avoid over-fertilization with nitrogen; vent and heat at dusk to reduce humidity.

Root and Stem Rots

Common hosts: All are susceptible.

Monitoring: Nutrient deficiency–like symptoms, usually on older foliage first; wilting and stunting. Media in containers may remain wet for long periods after watering.

Pythium: Infected roots are grayish-brown and appear water-soaked; root cortex easily pulls off when plant is pulled from soil; black stem. Favored by cool, wet soils, poor drainage, and soluble salts injury.

Rhizoctonia: Stems shrivel and turn brown at soil line. High humidity and warm temperatures favor infection.

Sclerotinia: Causes tan, brown or straw-colored stem cankers; black, hard masses of fungal growth called sclerotia (resemble rodent droppings) often form in pith or on stem. Under high humidity, cottony, white growth may appear on lower stem.

Sclerotium: Similar to *Sclerotinia*, but stems may become shredded. Sclerotia are small, brown, and resemble mustard seeds.

Management: Avoid overwatering or splashing soil when watering. Water early in day; keep water nozzles off ground. Remove and destroy symptomatic plants, test soluble salts levels regularly, and space plants to improve air circulation.

Powdery Mildew

Common hosts: Aquilegia, lupine, monarda, phlox, and many others.

Monitoring: Powdery white, grayish-white or brownish-white growth on upper and lower leaf surfaces or stems. Doesn't require free moisture on leaves to germinate and infect. Prolonged periods of high humidity and warm temperatures (above 70°F) favor development.

Management: Improve air circulation; use resistant cultivars.

Leaf Spots and Blights

Common hosts: Aquilegia, aster, *Bergenia*, chrysanthemum, coreopsis, delphinium, hosta, iris, *Liatris*, monarda, rudbeckia, and veronica.

Monitoring: Spots are discolored or dead areas on foliage or blossoms; blights result when spots coalesce or when large tissue areas are infected. May be spread through handling, through cuttings, and in seed. Splashing water spreads bacteria.

Management: Follow same procedures as for *Botrytis*. Avoid overhead watering and splashing water.

Viral Diseases

The most common viruses that cause diseases on perennials are: cucumber mosaic virus (CMV), tobacco mosaic virus (TMV), impatiens necrotic spot virus (INSV), tomato spotted wilt virus (TSWV), potyviruses (POTY), and alfalfa mosaic virus (AMV).

Monitoring: Black or brown spots on foliage; mosaic; stunting; black, brown, or chlorotic ring spots on foliage or stems; chlorotic or yellow oak-leaf pattern on leaves. Primary insect vectors are aphids and thrips (most commonly western flower thrips). Many viruses are also sap-transmitted: CMV, AMV and POTY (aphids); TMV (people); INSV, and TSWV (thrips).

Management: Inspect incoming plant material. Isolate suspect plants until diagnosis is confirmed. Destroy infected plants. Don't hold plant material year to year. Monitor insect populations. Grow resistant cultivars or virus-indexed material.

Rust Diseases

Common hosts: *Aconitum,* anemone, aquilegia, delphinium, *Liatris,* monarda, and potentilla.

Monitoring: Pale yellow or light green spots or pustules on upper leaf surfaces, powdery pustules on lower leaf surfaces (below chlorotic spots on upper leaf surfaces) or on petioles and stems. Pustules may be filled with rusty brown, yellow, cream, orange, white, or black spores.

Management: Water early in day; remove infected plants, bagging material to prevent spore dispersal.

Downy Mildew

Common hosts: Aster, artemisia, centaurea, geranium, geum, lupine, rudbeckia, and veronica.

Monitoring: Pale, yellow patches on upper leaf surfaces; older, infected areas may be browned; plants may be stunted. Under high humidity, undersides of yellow patches are covered with a dense, whitish, tan, gray, or violet/beige fungal growth. Older leaves are infected first. Can be seed-borne.

Management: Remove and destroy infected plants as soon as possible; remove all crop debris at end of season; remove related weed hosts; water early in day to promote drying before nightfall.

Culture Notes, June 1999.

Clean Stock: Your Plants Can't Always Live without It
Margery Daughtrey

Clean stock programs were first developed for carnations, mums, and geraniums during the 1950s and 1960s as a means of preventing catastrophic disease losses. Since that time, new clean stock programs have arisen to solve new challenges. Hiemalis begonias, New Guinea impatiens, double impatiens, angelonia, certain dahlias, and verbenas are more recent additions to the list of crops for which clean stock is available. Many additional crops could benefit from a carefully administered, clean stock program.

Why do we need clean stock? The alternatives to clean stock are the gruesome, difficult-to-control diseases that we plant pathologists counsel growers about regularly. Systemic diseases, in which the xylem or phloem of the plant is invaded, are the primary focus of clean stock programs. Once these truly dangerous pathogens are in your crop, there's very little that can be done to curb losses, other than discarding the affected plants to save the rest.

The Power of Prevention

Using pathogen exclusion to prevent disease, clean stock programs are disease management at its finest level. Exclusion is a critical management strategy in an industry that's primarily concerned with developing fetching new ornamental characteristics. Exclusion also works far better than after-the-fact treatment. The unfortunate truth is that chemical therapies aren't sufficiently effective as eradicants. We can minimize the spread of disease and we can minimize the impact of disease, but we can't, in most cases, cure a systemically infected plant with a pesticide.

The Development of Clean Stock

Luckily, there are some pathogen elimination tricks for those who want to offer their customers clean stock. Plants can be cleared of harmful bacteria and fungi through culture indexing. Growing plants at very high temperatures (near 100°F) for a few months and then propagating from the meristem tips can eliminate viruses. The resulting plants must be tested repeatedly to be sure that problematic viruses aren't present. Tissue culture without this follow-up virus indexing (using ELISA and/or indicator plants) is not sufficient.

The process of developing clean stock requires staff and facilities which allow a highly disciplined, precise program with a commitment to keeping the cleaned-up material clean. In addition, insect screening fine enough to exclude the tiny western flower thrips and positive pressure systems are used to help keep pests out of the greenhouse.

The Rewards

Remember the stunted, sickly New Guinea impatiens with black veins and dark spots that once were common? In the years since impatiens necrotic spot virus (INSV) and tomato spotted wilt virus (TSWV) first began to plague our industry in the mid-1980s, the dependability of the New Guinea impatiens crop has been assured by several companies' sincere efforts at clean stock programs. Other recent examples are clean stock improvements in some verbena lines, often contaminated by a dwarfing potyvirus, and in angelonia, a crop that has been crippled by a cucumber mosaic virus.

An important beneficiary of clean stock programs has been the florist's geranium, which owes its success largely to efforts against *Xanthomonas campestris* pv. *pelargonii* (Xcp), the agent of bacterial blight. Bacterial blight is highly contagious, and its leaf spotting and wilt are ugly and often fatal to the geranium—and to the grower's profits! A number of different techniques are used for ensuring Xcp-free geraniums. In the United States, the trusted technique is culture indexing, a system in which cuttings that have any bacterial or fungal contamination are discarded as unfit for elite mother stock. Several culture-indexing cycles and supplemental ELISA tests are used to ensure the cleanness of cuttings that pass the initial screen. Virus indexing is an equally important component of a clean stock geranium program because it means freedom from the symptoms viruses would otherwise cause, including ring spots, stunting, and leaf crinkling.

Chrysanthemums are another clean stock success story. Many techniques are employed: culture indexing, serological assays, and dot-blot hybridization, even grafting onto indicator cultivars that are very sensitive to one particular viroid disease. The program is designed to detect a wide range of diseases, including *Verticillium* and *Fusarium* wilt fungi and *Erwinia chrysanthemi* bacteria. Although heat treatment is used to free mum meristems from viruses, growth at a cool temperature (41°F) for eight to twenty-four weeks is needed for viroid elimination. A high-quality clean stock program for mums prevents plant and flower stunting, flower break, flower distortion, yellow spots, mottling, mosaic, ring spots, line patterns, or distortion of leaves, complete foliar chlorosis, stem cankers, dieback, wilt, and death.

Because much of the greenhouse flower industry depends on the efficient production of plants by vegetative propagation, clean stock is a necessity for reliable crop performance. Specialist propagators who can't deliver clean stock are easily outcompeted by those who can.

Pest Control, January 1999.

Demystifying Virus Indexing
Mike Klopmeyer

New and exciting flowering plants are the lifeblood of the floriculture industry. And some of the most interesting introductions are varieties that are vegetatively propagated. Cutting-propagated varieties allow breeders to bring unusual flower colors and plant forms to the market faster than they can bring seed varieties. The downside to this strategy is the threat of disease caused by bacteria, fungi, and viruses that can be transmitted in the cuttings.

Outside of the serious and well-recognized fungal and bacterial pathogens of cutting-propagated varieties (i.e., *Xanthomonas* bacterial blight of geranium), virus pathogens may be our biggest concern. Viruses can easily be spread by sap (plant to plant or people to plant) and insects such as thrips, aphids, and whiteflies. Breeders and propagators of cutting varieties have established certified clean plant programs that are designed to eliminate fungal, bacterial, and viral pathogens and then maintain these plants disease-free through the production cycle.

What Is Virus Indexing?
Virus indexing is a fancy word for thorough testing of the plant material for all known and unknown plant viruses. The two virus indexing types used are ELISA (enzyme-linked immunosorbent assay) and biological indicator plant inoculations. The ELISA technique is a low-cost, sensitive method that allows for relatively quick testing of a large number of plant samples using a

combination of antibodies that specifically react with certain viruses. Currently in the U.S., there are commercially available ELISA kits (Agdia Inc., Elkhart, Indiana) that target the major viral pathogens of floriculture crops. These kits are designed for laboratory use, but Agdia also markets easy-to-use on-site detection kits for tomato spotted wilt virus (TSWV) and impatiens necrotic spot virus (INSV).

Another sensitive virus detection technique is the use of biological indicator plants. Indicator plants are plant species that are naturally susceptible to many different plant pathogenic viruses. Leaves are removed from the plant to be tested, ground up in a special solution, then rubbed onto the leaf surface of the indicator plant. If viruses are present in the leaf sample, symptoms may appear on the indicator plant seven to twenty-one days after inoculation. This method of detection is more labor intensive and costly than ELISA is, but it provides a means to detect unknown viruses (where no ELISA kits are available). Reliable indicator plant species include various tobacco species, lamb's-quarter (a common weed), and petunia.

Eliminate Viruses

If virus indexing reveals the presence of viruses, infected plants need to be cleaned up. Elimination of viruses from infected plants is mainly accomplished through the use of heat therapy and special tissue culture techniques in the laboratory. These methods have been used through the years for vegetatively propagated crops including mums, geraniums, carnations, and potatoes. The method used is dependent on the types of the plant viruses that may infect the variety being cleaned. For example, for viruses that are unevenly distributed in the plant such as TSWV or INSV, the virus can be eliminated by removing the tiniest growing point, the meristem tip, and raising this plant artificially in a special tissue culture growth medium in the laboratory. Most viruses are incapable of advancing into the plant's youngest growing portion.

Viruses that are evenly distributed in the plant such as tobacco mosaic virus (TMV) or cucumber mosaic virus (CMV) are also very stable and easily spread. These viruses require additional treatments such as heat therapy (90° to 100°F) for up to one month, which will lower virus concentration. The combination of virus concentration reduction and meristem tip culture is very effective in eliminating the major viruses of floriculture crops.

After four to six weeks in culture, the tiny meristem tip (approximately 1 to 2 mm in size) develops into a small rooted plant that is taken into an insect-free greenhouse. Successful virus elimination from these plants needs to be verified by virus indexing. These clean plants can also be maintained in tissue culture in the laboratory to protect them from pathogens, insects, and other pests. They are only brought to the greenhouse for flower checks and buildup for production.

What Should I Do?

There is always a risk in bringing new varieties into your greenhouse. Make sure these plants are coming from a supplier who has a reliable clean plant production program from the lab all the way to the production greenhouses. Your duty is to keep these plants healthy until they are out the door. This includes an isolation area in the greenhouse to inspect all incoming plant material for insects and disease symptoms, a zero-tolerance program for virus transmitting insects such as thrips and aphids, and constant scouting of the crops for insects and disease and roguing symptomatic plants. When in doubt, send suspect plants for testing at your local university plant clinic.

Pest Control, May 1998.

Disease-Free Cyclamen

Hans S. Gerritsen

Traditionally, cyclamen has been a winter and spring plant. However, in some parts of the world, we see growers producing this crop year-round. In general, it's been considered to be a cool crop, but we have seen the nicest crops in the Sacramento Valley in California, where summer temperatures can exceed 100°F daily! Cyclamen performs in such conditions because of its native Mediterranean roots.

Common Disease and Insect Problems

While cyclamens are very sensitive to disease, they're very rewarding when grown under very strict clean conditions. *Fusarium oxysporum*, or *Fusarium* wilt, can be present latently within the plant and appear after transplanting or other stress periods, such as those from shipping, heat, or lack of moisture. Strict sanitation, clean soil, and clean pots are a must. Remove any yellowing plants. When cutting the corm of a diseased plant, a purple-red ring is visible. Many growers apply a preventative drench of Cleary's or Mycostop the fifth week of growing on.

Cryptocline cyclaminis is a wilt disease that creates stunted growth in the stems and leaves after transplanting. The fungus appears black at the edges. The vascular system will have reddish discoloration in petioles and the corm. Dead plant tissue will have masses of pale orange spores. The disease is spread rapidly by splashing water. Remove the diseased plants and apply Euparen.

Erwinia chrysanthemi, or soft rot, shows symptoms of sudden wilting followed by plant collapse. The corm appears slimy. Deep plantings plus warm, humid weather may trigger *Erwinia.* Cultural practices are the main problems—avoid overhead watering and high temperatures.

Botrytis cinerea is visible by its soft decay of the flowers and leaves. Flowers are spotted and have gray mold on them. Lower the humidity and increase the air circulation. Space crowded plants and avoid cold night temperatures. Some fungicides can be helpful, but they will be not needed if you make the greenhouse environment unfavorable for *Botrytis.*

Tomato spotted wilt virus (TSWV) shows yellow rings and brown streaks on the petioles. Flowers are malformed, and the plant stops growing. TSWV is transferred by thrips, and the only good cure is insect control.

Cyclamen mites create distorted leaves and curling, some discolored flowers, and twisted flower stems. Standard insect control will apply here.

Culture Notes, December 1998.

What's Wrong with Your New Guineas?

If you've wondered why your New Guinea impatiens wilt, stretch, burn, grow slowly, rot, or lack vigor, Jack Williams of the Paul Ecke Ranch offers some suggestions. Here are the most likely causes of ten of the most common New Guinea impatiens problems, as Jack presented them at the Southeast Greenhouse Conference in June. Use this handy list to diagnose your next New Guinea problem:

Problem 1: Cuttings rot during propagation
- Too wet
- Too cold

- Unsterilized media
- Damaged cuttings
- Disease

Problem 2: Cuttings fail to develop roots
- Media too cold
- Too wet
- Too dark

Problem 3: Cuttings or young plants stretch or elongate
- Cuttings are overcrowded
- Too dark
- You didn't transplant on time

Problem 4: Plants are slow to take off and establish
- Greenhouse too cold
- Soil too moist
- Excess fertilizer

Problem 5: Leaf tips burn and leaves curl
- Excessive drying of media
- Excessive levels of fertilization

Problem 6: Leaves turn chlorotic (yellow) along edges and plants lack vigor
- Fertilizer required
- Florel treatments
- Root rot

Problem 7: Plants show symptoms of insect damage
- Aphids
- Mites
- Thrips

Problem 8: Plants rot and die for no apparent reason
- *Botrytis*
- *Rhizoctonia*
- Virus (TSWV/INSV)

Problem 9: Plants are adequately sized but fail to flower
- Too dark
- Too hot

- Too much fertilizer
- Florel

Problem 10: Plants rapidly die in the consumer environment

- Plants are too "soft"
- Inappropriate conditions
- Cultivar selection

Culture Notes, August 1999.

Chapter 8
Specific Diseases
Bacteria

Protect Your Geraniums from *Xanthomonas*

Geranium season—and your greenhouses—will soon be heating up, and so will the risk of a *Xanthomonas* outbreak. While this is an aggressive disease at best, successful practical control is possible by taking a few precautions and knowing how to control it.

Sanitation

Disinfect all greenhouse benches that held geraniums in the past season, along with any equipment used to transport, hold, or work with geraniums. Sterilize planting containers, cutting knives, and associated greenhouse implements. Use clean growing media free from debris.

Clean Cuttings

Take cuttings from culture-indexed plants. Inspect plants from other sources before allowing them in your greenhouse. Avoid overhead watering. Space plants to avoid plant-to-plant contact, and reduce splashing water from plant to plant.

Rogue and Spray

Roguing, or removing infected plants, is your first line of defense if you find *Xanthomonas* in your greenhouse. When removing visibly diseased plants, be careful not to spread the disease with your hands during the process. When practical, isolate particular varieties or groups of plants that are showing more symptoms than others are.

Research at The Ohio State University, Columbus, has uncovered an interesting way that you may be able to get ahead of the disease when producing geraniums in hot weather. Pathologist Steve Nameth found that plants turned yellow after spraying with the bactericide Phyton-27 and that the yellowing seemed to be in areas highly infected with disease. In his experiments, the leaves of infected plants turned bright yellow within five days of spraying. Classic *Xanthomonas* symptoms were seen both on inoculated plants that were sprayed and that yellowed and on inoculated plants that weren't sprayed. Symptoms on unsprayed plants didn't appear until more than one week after the yellowing appeared on the infected, sprayed geraniums.

Geranium stock plant growers at many locations observed this yellowing effect after spraying young, rapidly growing stock plants with Phyton-27 during the hot summer months, even when the plants weren't showing symptoms of *Xanthomonas*. This yellowing phenomenon hasn't been reported in mature stock plants during fall and winter.

For this technique, Source Technology Biologicals Inc., manufacturer of Phyton-27, recommends spraying 5 oz. per 10 gal. on geraniums, with the exception of ivy geraniums, which you should spray with 3.5 oz. per 10 gals.

In lab tests, the yellowed foliage was consistently tested positive for the presence of *Xanthomonas,* but absence of yellowing doesn't mean your crop has no risk of disease outbreak.

Culture Notes, January 1999.

Bioprotecting Anthurium against Bacterial Blight

Anne Alvarez, Ryo Fukui, Hisae Fukui, and Rosemary McElhaney

The expanding anthurium industry still fails to reach its potential because of losses due to anthurium blight, caused by the bacterial pathogen *Xanthomonas campestris* pv. *dieffenbachiae* (Xcd). First reported in Hawaii, the disease has since been widely observed throughout the tropics and subtropics, including such diverse production areas as Jamaica, Taiwan, and the Philippines. Recent reports indicate that bacterial blight has been detected in India and in the Netherlands. Antibiotic sprays, first used to combat blight, were abandoned in Hawaii in the early 1990s because of their high cost and failure to eliminate antibiotic-resistant Xcd strains. Control by strict sanitation and protection with selected pesticides has reduced disease spread, but these measures alone are inadequate. Most anthurium cultivars are susceptible to blight, and in a market that demands a wide variety of shapes, sizes, and colors, resistance breeding is exceedingly slow. Bacterial blight hasn't been eradicated from production fields because symptomless (latent) infections persist without detection in plant tissues and perpetuate the disease in symptomless plants.

A novel approach to controlling this devastating disease has been the use of protective sprays containing mixtures of bacteria that inhibit infection and spread of the pathogen. Ryo Fukui and Hisae Fukui, working in our laboratory at the University of Hawaii, Honolulu, discovered that anthurium can be protected from infection by spraying leaves with a mixture of living, nonpathogenic, bacterial antagonists that serve as biological control agents (BCAs).

In a search for the most appropriate BCAs, they compared growth of various bacteria in fluids taken directly from anthurium plants. Some bacteria isolated from anthurium inhibited the multiplication of the blight pathogen. The researchers found that certain mixtures of four to five bacteria were far more effective in reducing disease than were applications of a single species. This form of biological control has been very promising in twenty-five separate greenhouse trials and two field trials in a semiprotected shade house, where anthurium are commercially grown in the tropics. The BCAs' effectiveness isn't reduced by many of the common fungicides and insecticides routinely applied on anthurium.

To achieve lasting control in the field, we need research underlying the basic mechanisms of biological control. The challenge will be to provide a suitable formulation of BCAs, as well as cost-effective methods for lasting protection under field conditions. We're now optimizing the quantities and timing of protective sprays and drenches for transplants and investigating ways to establish BCAs on or within plant tissues as they are acclimatized from in-vitro cultures.

Production of pathogen-free (tissue-cultured, triple-indexed) planting stock, followed by bioprotection with BCAs, will form the basis of long-range control in conjunction with current practices of selected bactericide applications, sanitation, and cultural disease control methods. If this approach is successful with anthurium, a new integrated plant propagation strategy could be developed for protecting other plants against bacterial diseases.

The treatments on anthurium are used as a model system for other ornamental aroids (*Aglaonema, Spathiphyllum,* philodendron, etc.) that are affected by the same disease. In-vitro grown anthurium are usually pathogen-free as they emerge from flasks, but they have no protective microflora. Young plantlets that are misted on benches as they're established in trays and pots are particularly vulnerable to infection.

The focus of current research is to extend the biological control applications to tissue-cultured plants. We aim to develop appropriate balanced formulations for inoculating plants as they're deflasked, acclimatized, and established in pots. The potted plants eventually should have the necessary microflora for bioprotection against disease. If this strategy for bioprotection proves successful, it could prompt a major change in production procedures for potted ornamental aroids.

Using a Bioluminescent Strain to Monitor Infection

Other plant pathologists have used bioluminescence to trace the infection process of a related bacterial pathogen, *X. campestris* pv. *campestris*. A bioengineered strain containing the "lux" gene was successfully used to monitor black rot disease spread in cabbage fields. In our laboratory, Rosemary McElhaney bioengineered a strain of the pathogen Xcd by transferring the lux gene from another bacterium (a strain of *Escherichia coli*) making the pathogenic Xcd strain bioluminescent without changing any of its other characteristics. We have used this property to detect the pathogen in plant tissues and monitor the internal infection process without relying on dissection, culture, or visual symptoms.

How does it work?

A highly sensitive X-ray film is placed over the plant tissues, and the film is exposed by light emitted from the bioluminescent bacteria within the infected areas. Hence, we can "photograph" disease progress. The method is nondestructive and allows symptomless infections as the plant grows to be seen.

We used the bioengineered strain to evaluate susceptibility and resistance in thirteen different anthurium cultivars and to determine the extent of systemic movement in the plant with different temperatures and nutrition. We also did experiments to determine the compatibility of BCAs with pesticides commonly used in field operations to control fungal pathogens and insects. So far, we've found few pesticides that are harmful to BCAs.

Making it work in the field

The last aspect of ongoing research involves an attempt to produce BCA inoculum in a readily applied form for large-scale field applications. Results with freeze-dried preparations of inoculum showed that this method of delivery is effective, but further research is needed to determine the optimum conditions for large-scale growth of BCAs and to evaluate the most effective cryopreservants to freeze dry the BCAs.

It isn't easy to produce the same biological control observed under carefully controlled greenhouse conditions in the field. Success with biological control of anthurium blight in the greenhouse and field is determined by the following: (1) The antagonistic bacteria significantly reduce infection by Xcd, provided that inoculum is introduced as a mixture of four bacterial species, not as a single component. (2) Inoculum must be applied and maintained in high concentrations to achieve control. (3) Many commonly used pesticides don't interfere with biological control. (4) The BCAs appear to work by out-competing the pathogen for nutrients rather than killing it directly. (5) The competitive advantage of BCAs is enhanced by specific nutrients.

Anticipated future benefits

Growers could fill an important niche by producing disease-free, tissue-cultured and bioprotected ornamental aroids for the market. Based on promising results of previous experiments, BCAs should protect young plants from disease. Understanding biocontrol using mixtures of bacterial antagonists should help establish basic principles for BCA applications that can be extended to control bacterial diseases in other potted ornamental aroids.

February 1999.

Botrytis Research

Don Grey

Growers may have a new biological pesticide to fight *Botrytis*. This product, known under the test name of FP-8, is a bacterium in the Bacillus family. The biocontrol agent is undergoing research at the USDA Agricultural Research Service's Horticultural Crops Research Lab in Corvallis, Oregon. Research horticulturist Walt Mahaffee and his team have been using FP-8 on geranium cuttings; other crops will also be tested. Results appear promising. As a biocontrol agent, FP-8 is expected to have a lower reentry interval and may offer fewer compliance issues with the EPA's Food Quality Protection Act.

Culture Notes, March 1999.

Bacterial Blight of Geranium

Wendy O'Donovan

Bacterial blight is a terminal disease, no ands, ifs, or buts. Seed, ivy, and cutting geraniums are all susceptible to this disease. Regal geraniums and perennial geraniums, although seemingly unaffected, can be carriers. The agent, *Xanthomonas campestris* pv. *pelargonii*, is a vascular bacterium that progressively colonizes the water-conducting tissues, causing plant death. Symptoms vary according to many factors including the cultivar or species infected, mode of infection, greenhouse environment, and conditions under which the plant was grown.

Symptoms

When seed or cutting geraniums are infected by water splashing or dripping from infected plants, or by mist in propagation, a leaf-spot stage develops. Small sixteenth- to eighth-inch water-soaked spots appear on the underside of the leaves and become well defined in just a few days. Then they are more visible as tan to brown spots approximately a tenth-inch in diameter on the upper side of the leaves. The bacteria infect the entire plant by spreading through the leaf petioles into the plant stem via the water-conducting tissue. The lower leaves may yellow, the plant will wilt, and eventually it will die.

Systemically infected plants do not exhibit the leaf-spot stage. These plants do not grow with the vigor of healthy plants, wilting may occur, and the stem of the plant turns black. Ivy geraniums behave differently. They never wilt, but lack vigor and may show edema-like symptoms or look as though they have a mite infestation. Their leaves dry up, remaining attached to the plant.

Under cool, dark conditions, infected plants may not exhibit obvious symptoms for some time, being insidious while allowing for the infection of other plants. The bacteria are spread by leaf-to-leaf contact, on the hands and clothing of employees, and on tools such as knives when taking cuttings or when spacing or pinching plants. Rapid spread can occur by way of capillary mats and bench-flooding watering systems.

Cuttings infected with *X. pelargonii* root more slowly than healthy cuttings. In the propagation bed, the bacteria are released from the vascular system of infected cuttings and they spread freely to infect other cuttings.

Losses, such as rotting from the base of the cutting upward, will be observed, although not all infected cuttings may die. Sometimes infected cuttings will root in propagation and show no apparent symptoms.

Because the symptoms of this disease are so varied and because other pathogens cause similar symptoms, diagnosis should never be based on observed symptoms alone. A comprehensive laboratory diagnosis is necessary.

Control and Prevention

No chemical sprays will cure plants of bacterial blight, and those that suppress bacterial spread promote false security. Control of *X. pelargonii* is best accomplished by avoidance. Purchase only culture-indexed cuttings from a reputable producer. Using only one source when purchasing your geraniums is best. If multiple sources are used, keep the plants isolated from each other and isolate seed geraniums from other geraniums. Institute a first-class sanitation program before your plants arrive and continue it throughout the life of the crop.

Sanitation

Whether the disease has ever been in your greenhouses or not, the best way to begin the geranium season is clean. Because blight-causing bacteria persist in plants and plant debris, discard all geranium plants and thoroughly remove all plant debris. Haul all plants, debris, and soil away where no one can visit them in a compost heap and return to your greenhouse carrying the pathogen. Greenhouse walls and floors, benches, watering systems, and hoses should be sanitized. Use new flats, pots, capillary mats, and sterilized media to eliminate any possibility of bacterial introduction.

The most important jobs ahead are to educate your personnel concerning sanitation procedures and to work together to keep the disease out. Be aware that the bacteria can successfully hitchhike on delivery trucks, racks, tools, and clothing. Garden center and outdoor geraniums can be a source of infection and should never be handled before working with the geraniums in growing areas. Enter the greenhouse area each day with freshly laundered clothing, and wash your hands before entering. Avoid unnecessary handling of geranium plants, and wash your hands frequently. Restrict the traffic flow to avoid movement through the geranium area. All tools, delivery racks, knives, garbage cans—any items that come into the greenhouse area—are suspicious and must be sanitized. Never hang ivy geraniums overhead where the irrigation water can drip onto other geraniums, and do not use overhead irrigation. If an infection should occur, early detection and a strong sanitation program go a long way to limit the spread of the disease.

Bacterial blight is a serious disease; however, culture-indexed cuttings are the foundation of a grower's blight-prevention program. A well-informed grower using carefully executed sanitation procedures can produce beautiful, blight-free geraniums year after year.

Pest Control, December 1997.

Southern Bacterial Wilt Found in Geraniums

P. Allen Hammer and Karen K. Rane

Most geranium growers are familiar (maybe too familiar) with bacterial blight caused by *Xanthomonas* (*Xanthomonas campestris* pv. *pelargonii* [Xcp]). It's the most serious infectious disease of geraniums.

This year another bacterial wilt disease has been found in zonal geraniums. *Ralstonia solanacearum* (Rs), formerly called *Pseudomonas solanacearum,* is the bacterial pathogen that causes the disease known as southern bacterial wilt.

Symptoms of Xcp include leaf spots, wilt and death of zonal, ivy, and seed geraniums. The disease is usually introduced into a greenhouse through infected cuttings, dried leaves, or other residue from infected plants, or by moving outdoor plants into the greenhouse. The bacterium also causes leaf spots on perennial cranesbills (*Geranium* spp.), which can serve as a source of disease for greenhouse pelargoniums if growers produce both annuals and perennials.

Southern bacterial wilt's symptoms resemble those of the wilt phase of bacterial blight caused by Xcp—wilted leaves, yellowing, eventual death of plants.

But unlike bacterial blight caused by Xcp, there are no discrete leaf spot symptoms with southern bacterial wilt. Rs invades plants primarily through the roots, and it can be introduced to a greenhouse crop through all the same methods as bacterial blight. Both pathogens invade the vascular system of the plant, block water transport, and result in wilt and death.

Which Disease Is It?

Why is it important to determine which bacterial wilt disease is responsible for wilting geraniums when both diseases are fatal and the pathogens spread in a similar manner? There are two major differences in these pathogens, and they have a potentially significant impact on disease management. The first difference is host range. While Xcp infects only *Pelargonium* and *Geranium* species, Rs is reported to have a wide host range that includes numerous vegetable and herbaceous ornamental crops. It's possible for Rs to spread to other crops in a greenhouse, especially if irrigation water from infected geraniums runs down the bench and comes in contact with roots of other plants.

The second difference is pathogen survival. Xcp cannot survive for long periods of time outside of host tissue. When Xcp-infected plant debris decays, the pathogen population declines dramatically. This is why clearing a greenhouse of all geranium plants and plant debris effectively reduces the chances of a recurrence of the disease to near zero. In contrast, Rs is known to survive for several years in the soil of warm climates without the presence of susceptible host tissue. Severe disease outbreaks have occurred in susceptible field crops after long rotation with non-host crops. This means that the potential exists for Rs to survive in greenhouses on soil floors or benches with contaminated soil, sand, or other substrates.

Special laboratory tests are required to distinguish between Xcp and Rs. Growers should send any wilted geranium plant to a diagnostic laboratory for confirmation of bacterial disease. Diagnostic laboratories are also able to identify other causes of wilt symptoms, such as root rot diseases and high fertilizer salts. It's always wise to determine exactly what you are dealing with so you can make proper management decisions.

Preventing Further Infection

Sanitation is still the most important cultural practice for managing both Xcp and Rs. All symptomatic plants, pots, and soil should be discarded. Healthy-looking plants next to wilted ones are often discarded because these plants may eventually also develop symptoms. Many growers place plants into large garbage bags as they move through the greenhouse bench by bench, reducing the chances that infected leaves or infested growing mix will drop to the floor as they carry plants out of the greenhouse. After an outbreak of Rs, bench surfaces should be disinfested with a quaternary ammonium compound before introducing the next crop. This is also a good practice after an Xcp outbreak if geraniums will be grown as the next crop.

Spray applications of copper compounds may help protect uninfected plant surfaces from Xcp infection, but they won't cure diseased plants. If possible, avoid hanging ivy geraniums over other geraniums on benches, as bacterial blight symptoms in ivy geraniums are often less distinct and the pathogen can be spread by dripping water to the crop below. Southern bacterial wilt in ivy geraniums hasn't yet been diagnosed, but the plant may be a host for Rs. Additional management practices will depend on the specific bacterial disease as well as the individual characteristics of each greenhouse system. Bacterial wilt diseases in geranium don't necessarily result in total crop loss. A quick and accurate assessment of the problem followed by the use of effective management practices can reduce losses due to any infectious disease, including bacterial diseases of geranium.

July 1999.

Bacteria versus Geraniums

Margery Daughtrey and Karen K. Rane

Xanthomonas

Geranium trouble for 1999 began brewing back in fall 1998. Some newly established geranium stock plants started to show leaf spots or wilting, and the dreaded diagnosis of bacterial blight or *Xanthomonas* began to be heard across the U.S. and Canada. Recently, a single report of this same disease in July (Steve Nameth, *Ohio Hotline,* July 1999) was warning enough that growers would need to be on the lookout for bacterial blight in their year 2000 crop. Other cases reported since July have made it obvious that *Xanthomonas campestris* pv. *pelargonii* is going to be a player in the geranium game in the new millennium.

Ralstonia (Pseudomonas)

A second perpetrator of systemic bacterial disease trouble was diagnosed beginning in January 1999. Although the symptoms were quite similar, the pathogen had a different name: *Ralstonia solanacearum,* or, as it was called until fairly recently, *Pseudomonas solanacearum.* This bacterial disease, called southern wilt, wasn't new to science, but it was new to most geranium growers. They watched as some of their plants wilted down, often showing wedges of yellowing or browning of the leaves. The only difference between bacterial blight and southern wilt growers could see was the continued absence of any round leaf spots when southern wilt was the problem.

Currently there are no indications that southern wilt will be returning to haunt growers next spring. Because of this pathogen's known ability to affect a wide range of plants and to persist in fields from year to year, growers who were troubled by this problem in 1999 will keep their fingers crossed next season. To the best of our knowledge, greenhouse operations where the disease was detected this spring didn't experience spread of the bacteria into other crops during the course of the season.

Assorted Leaf Spots

Although they weren't common in 1999, there are several other, usually minor bacterial diseases that can affect geraniums. Only when environmental conditions come together to favor disease development will the leaf spotting be widespread and alarming.

The bacteria most often implicated in these leaf spot outbreaks are *Acidovorax* sp., *Pseudomonas cichorii*, and *P. syringae*. Seed geraniums are

Key Disease Minimization Strategies

- Keep geraniums from different suppliers separate.
- Don't grow hardy *Geranium* spp. near greenhouse *Pelargonium* spp.
- Separate seedling geraniums from the cutting types (which are more likely to be systemically infected).
- Never grow ivy geraniums in hanging baskets above any other seedling or cutting geranium crop.
- Use drip irrigation systems on stock plants to eliminate spread of bacteria by splashing.
- Take time out for a sanitation step between cultivars when you're taking cuttings.
- Isolate your stock area.

more commonly affected than vegetative geraniums, but both are potential hosts for these bacteria. One of the most important aspects of these nonsystemic, leaf spot–only diseases is that they add to the impossibility of making a reliable sight diagnosis of bacterial blight (*X. pelargonii*). If leaf spots are the only symptoms present, the disease could easily be caused by another bacterium that's much more manageable than *Xanthomonas*. Laboratory testing is necessary to discover exactly which bacterium or fungus is responsible for leaf spots.

Bacterial Disease Management

Whatever the bacterial disease, impeccable sanitation and a scout with a good eye are the best defenses. Early disease detection is essential for keeping losses to a minimum. With generous roguing out of the symptomatic plants and those nearby, you have a good chance of avoiding a major crop loss. Copper sprays may also be helpful for reducing the risk of disease spread within the crop. The growers who were faced with *X. pelargonii* or *R. solanacearum* contamination last season are to be commended for their alertness and prompt responses—they discarded the infected and suspect plants and got on with the business of finishing a healthy crop. Growers who learn to take all of the appropriate precautions on a regular basis will be able to weather the storm of a bacterial invasion.

Technology Sources for Bacterial Identification

Agdia Inc., Elkhart, Indiana—Supplies labs with multiwell ELISA equipment for identification of *Ralstonia solanacearum* and *Xanthomonas pelargonii*. Accepts plant samples from growers for diagnosis. Tel: 800-622-4342.

BIOLOG, Hayward, California—Supplies labs with carbon utilization tests and software for bacterial ID. Does not accept plant samples.

Microbial ID Inc., Newark, Delaware—Supplies labs with fatty acid analysis for pure cultures of bacteria. Does not accept plant samples.

Neogen Corporation, Lansing, Michigan—Supplies labs with multiwell ELISA equipment for *Xanthomonas pelargonii* identification. Does not accept plant samples.

Pest Control, October 1999.

Bacterium Keeps Poinsettias Short and Lovely

A bacterium without cell walls is behind the poinsettia plant's beauty and popularity, according to researchers. Studies show that the bacterium without walls, called a phytoplasma, serves as a dwarfing agent that keeps the poinsettia from growing to eight feet or more, as it does in its native home of Mexico.

The phytoplasma triggers a hormonal imbalance that instructs the plant to form more auxiliary branches, so it grows outward instead of up. This phenomenon, called free branching, also produces more of the colorful bracts.

Until recently, scientists suspected the poinsettia mosaic virus as a causative agent. U.S. Agricultural Research Service scientists and personnel at Ball FloraPlant, West Chicago, Illinois, exposed the mosaic virus for what it is: a costly nuisance that plays no part in free branching.

Culture Notes, March 1999.

Fungi

Anthracnose Continues to Damage Fern Crops

Anthracnose disease, long a problem in Central America, has been leaving its mark on Florida ferneries, and researchers offer little assurance that the problem will end soon. The fungus, which began attacking U.S. operations in 1992, probably came from Central America, as Panama, Guatemala, Costa Rica, and El Salvador all report disease problems, Robert Stamps, professor of environmental horticulture and extension cut foliage specialist at the University of Florida's Apopka research station, says in *Flower News.*

In the last four years, anthracnose has destroyed about 30% of many growers' crops. It attacks leatherleaf ferns, infecting new cuticles in young leaves that are unable to fight disease. They become brown or black and can't be harvested. Though plants are safe when they reach stages six and seven unaffected, when one plant is infected, spores spread quickly and can destroy crops.

Keys to containing the fungus:
- Don't mow or burn to destroy hot spots. This can spread spores easily.
- Improve ventilation systems to prevent stagnant air.
- Avoid overwatering and use dry fertilizers rather than liquid.
- Make sure all employees disinfect themselves when moving between ferneries.
- Fungicides with the active ingredients mancozeb, chlorothalonil, and propiconazole are helpful, but nothing can cure the disease. Frequent fungicide applications (a two-per-week cycle) are essential to prevent reintroducing the fungus.

Culture Notes, February 1997.

Controlling *Cylindrocladium* on Spaths

A. R. Chase

Spathiphyllum, sometimes called peace lilies, are tropical, perennial plants grown in various sizes, ranging from dish gardens to large specimens to mass plantings in malls and office buildings. Compared to other aroids such as *Dieffenbachia* and *Aglaonema,* few diseases occur on this plant. The most important diseases are caused by fungi including cylindrocladium root and petiole rot, *Erwinia* blight, *Myrothecium* petiole rot, and *Phytophthora* aerial blight.

Far and away the most serious disease of *Spathiphyllum* is *Cylindrocladium* root and petiole rot caused by *Cylindrocladium spathiphylli.* This disease was discovered in the early 1980s and rapidly spread throughout all states and countries producing *Spathiphyllum.* One of the first symptoms of the disease is yellowing of lower leaves, sometimes accompanied by slight wilting. Elliptical dark brown spots may be found on leaves and petioles; lower portions of petioles frequently rot. At this stage, plant roots are severely rotted, and few roots are healthy. Tops of such plants are easily removed from the pot without any adhering roots.

In the 1980s, extensive testing was performed to determine possible controls for this disease, evaluating fertility, temperature, potting medium, cultivar, and fungicides. Overall, cultural controls that kept the disease out of the nursery were the only reliable methods for controlling this devastating disease. These include pathogen-free plants from tissue culture or seed sources, using sterilized potting medium and pots, and growing plants on clean or disinfected raised benches. The *Spathiphyllum* cultivars tested to date have been very susceptible to *Cylindrocladium* root rot, with the exception of *S. floribundum.* This species is a host of the pathogen, but is highly resistant and shows little root loss when infected with *C. spathiphylli.*

Chemical treatments haven't been completely effective unless disease pressure is low. Under conditions of high disease pressure, especially during the summer, triflumizole (Terraguard from Uniroyal) has provided better control than available alternatives. In 1996, the Central Florida Research and Education Center in Apopka reported an efficacy trial using certain fungicides not previously tested for *Cylindrocladium* control on *Spathiphyllum.* Results indicated that fluazinam (an experimental compound from ISK Biotech—recently purchased by Zeneca), copper pentahydrate (Phyton-27 from Source Technology Biologicals), and thiophanate methyl (Cleary's 3336 from W. A. Cleary Corporation) each showed a significant degree of efficacy.

The continuing need to evaluate new products for *Spathiphyllum* diseases was recognized by Florida *Spathiphyllum* producers, and a contract was set up with Chase Research Gardens Inc. During the past year, we have run tests on a wide range of potential controls including biologicals, broad-spectrum chemicals, and combinations of products.

Performance Results

Seven trials were performed using various rates of seven products for control of *Cylindrocladium* on *Spathiphyllum*. Some products are currently available for use on ornamentals, while others are under development. All companies have expressed a continuing interest in labeling effective products for use on ornamentals. Results varied from test to test, based on disease severity, which was in turn dependent on temperature. During the summer, when temperatures were the highest, disease severity was also the highest.

Summary of Seven Trials Performed on Seven Fungicides for Control of *Cylindrocladium* Root and Petiole Rot on *Spathiphyllum* 'Petite'.

Product	Availability	Manufacturer	Rate/100 gal.	Number of times tested	Results
Cleary's 3336 50WP	Yes	W. A. Cleary Chemical Corp.	8 to 32 oz.	Three	Very good to excellent
Decree 50WP	No	SePRO Corp.	8 to 32 oz.	Three	Very good
Heritage 50WP	No	Zeneca Ag Products	1 to 8 oz.	Two	Good
Medallion 50WP	Yes	Novartis Crop Protection	1 to 4 oz.	Six	Very good to excellent
Phyton 27	Yes	Source Technology Biologicals, Inc.	8 to 35 oz.	Three	Poor to very good
Rootshield Biological Fungicide	Yes	BioWorks	8 to 16 oz.	Four	Poor to very good
Terraguard 50WP	Yes	Uniroyal Chemical Company Inc.	4 to 8 oz.	Five	Poor to good

Products marketed on ornamentals at present are labeled "yes." Those not marketed at present are labeled "no." Be sure to check product labels for specific uses on *Spathiphyllum* for *Cylindrocladium* root and petiole rot. Follow rates and intervals according to manufacturer's directions and not experimental protocols.

Some products (such as Phyton-27 and RootShield) performed well under low to moderate disease pressure, but not as well under high disease pressure. Decree 50WP (SP2001, figs. 1 and 3) gave good control when used at 32 oz. per 100 gal. and may be useful in a rotation with other products when disease pressure isn't at its highest. Similarly, Heritage 50WP gives good control when used at 1 to 8 oz. per 100 gal. Cleary's 3336 50WP (8 to 32 oz. per 100 gal.) and Medallion 50WP (1 to 4 oz. per 100 gal.) produced the highest quality plants in trials where they were used.

Terraguard 50WP continues to give good disease control when used at rates of 4 to 8 oz. per 100 gal. as a weekly drench. Mixtures of Terraguard and Medallion didn't perform better than the products used alone. Medallion and Heritage performed the best under all levels of disease pressure and perhaps should be saved for conditions when other products may fail. Phyton-27 worked well when used at 32 to 35 oz. per 100 gal., with lower levels of control seen when lower rates were employed.

Finally, RootShield worked well at 8 oz., but had a lower level of control when used at higher rates. This may be due to direct phytotoxicity to *Spathiphyllum* roots or perhaps to interference with other soil microorganisms that aid in reducing activity of *Cylindrocladium spathiphylli*. Note that the RootShield label doesn't call for weekly applications of this product.

Controlling *Cylindrocladium* root and petiole rot on *Spathiphyllum* must still be based on good cultural practices and especially on the use of pathogen-free plants. The spectrum of effective fungicides for control of this disease is greater now than it has been at any time during the past fifteen years. Wise use of a variety of products, both chemical and biological, should lead to the best control programs for most growers.

Editor's note: The research reported here was funded by the National Foliage Foundation, Chase Research Gardens Inc., SePRO Corporation, Source Technology Biologicals Inc., and Zeneca Ag Products. Bradford Botanicals and Twyford Plant Laboratories supplied all plant materials required.

Pest Control, August 1998.

Restraining Rusts on Your Crops

Margery Daughtrey

What do iron frying pans, geraniums, fuchsias, and snapdragons have in common? Susceptibility to rust! Although the iron in your pipes won't attack your plants, a contagious plant disease called rust will. When a rust fungus attacks, pale spots usually appear on upper leaf surfaces opposite distinctive pustules (bumps) on leaf undersurfaces that open to release masses of spores that look like colorful dust. The diseases are called rust because often this dust is reddish brown, but spores may also be white, yellow-orange, dark brown, or black.

Because rust spores are spread great distances on wind currents, infected ornamental plants or weeds may be a source of spores to initiate epidemics during crop production. Although rusts are very host-specific (able to attack only a few species), sometimes flower crops have a close relative in the local flora that may share their susceptibility to a rust fungus. In this way, fireweed may supply spores for fuchsia rust outbreaks, just as common mallow or other *Malva* species may provide rust spores for hollyhocks. There is less likelihood of weeds serving as the source of rust spores for outbreaks of geranium or snapdragon rust. Rust spores may be on plants you purchase, but symptoms may be delayed until a wetness period allows them to infect.

Each year at least a few cases of geranium rust (*Puccinia pelargonii-zonalis*) occur in the geranium trade. Cuttings are often produced in areas where geraniums are able to grow year-round, and rust spores may occasionally be conveyed from infected geraniums by air currents. The initial symptoms of geranium rust are tiny, yellow spots on upper leaf surfaces that begin to show seven to ten days after infection. If temperature conditions are right for disease development, a small pustule will then begin to form opposite the yellow spot on the leaf undersurface, and a pile of cinnamon-brown spores will erupt from this pustule. One or more rings of pustules may develop around the first ones, and then the disease is very easy to identify.

The florist's geranium, *Pelargonium* x *hortorum*, is by far the most susceptible host for geranium rust. Some cultivars develop hundreds of pustules per leaf. Rust is more often seen on cutting geraniums, but seedlings are also susceptible. Scented, regal ('Martha Washington') and ivy geraniums don't get rust. Once rust is within a production facility, it's difficult to eradicate if geraniums are grown continuously, because spores can survive for up to three months outside of the plant. Infections will occur readily when leaf wetness lasts five to six hours and temperatures are 54° to 75°F. These conditions allow spores to germinate and penetrate through stomates on either leaf surface.

At an optimum temperature of 70°F, a new supply of spores will be ready in fourteen to nineteen days. To break the cycle of geranium rust, keep your greenhouses free of geraniums for at least three months.

Fuchsia rust (*Pucciniastrum epilobii*) has a true "alternate host." Certain fir species host a stage of the same rust fungus that affects *Epilobium* (fireweed) and fuchsia during the other part of its life cycle. Fuchsia rust causes large, patchy leaf spots with purple rims; masses of yellow orange spores are produced on leaf undersurfaces. This disease is most destructive during propagation.

Snapdragon rust (*Puccinia antirrhini*) is more of a problem for outdoor cut flower growers than for bedding plant producers. Large quantities of dark, chocolate-brown spores are produced on foliage and stems. Germination and infection can take place from roughly 45° to 65°F, as long as leaves are wet for six to eight hours. Temperatures higher than 70°F curtail infection. Because cool conditions foster infection and warmer conditions speed up spore production, temperatures between 55° and 70°F are ideal for epidemic development. New spore production takes as little as nine days at 70°F. High temperatures (80° to 90°F) abort young infections and halt disease development.

Rust management is achieved with a combination of cultural controls (keeping leaf surfaces dry) and chemical control (alternating protectant and systemic fungicides). In rare instances, controlling particular weeds in the vicinity may help to reduce a rust inoculum source.

Fungicides[1] effective against rust include protectants such as sulfur, mancozeb (Protect T/O, Dithane) and chlorothalonil (Daconil, Exotherm), and systemics such as triadimefon (Strike), myclobutanil (Systhane) and oxycarboxin (Plantvax). For outdoor crops, the systemic propiconazole (Banner) is another option. Scout regularly for rust symptoms on susceptible crops—early detection is important.

[1]Trade names are used for convenience only. No endorsement of products is intended, and no criticism of unnamed products is implied. Follow label instructions.

Pest Control, July 1997.

Geranium Rust in the New Century

A. R. Chase

Geraniums are subject to many diseases including *Botrytis* blight, *Xanthomonas* blight, *Alternaria* leaf spot, and several viruses. Periodically they've been severely damaged by rust disease as well.

Rust on zonal geraniums is caused by *Puccinia pelargonii-zonalis*. It appears to have been discovered in South Africa, where the plant was discovered and collected. The disease spread with the geranium's introduction throughout the Southern Hemisphere, but it didn't become established in the Northern Hemisphere until the 1960s. At that time, it spread rapidly and was found throughout the U.S. by the middle of the 1970s. It's interesting that the outbreak in the 1970s was controlled through a variety of cultural and chemical means, but the disease wasn't immediately recognized when it appeared again in the middle of the 1990s. We experienced about twenty years without significant rust problems on this crop, so today's growers may not have ever seen the disease on their geraniums.

Symptoms and Susceptibility

Geranium rust symptoms are typical of most rust diseases. Pale green or yellow spots show on the upper leaf surface followed by the development of a dead area in their centers. The pustules form first on the leaf undersides, where they erupt into the reddish-brown rust color that gave the disease its common name. These pustules (raised blisterlike areas) are often single but may be surrounded by a ring of smaller pustules due to new infections. The spores are easily spread by wind or fans, as well as handling by workers. The spores will infect new plants or new leaves if they land on them when plants are wet. Leaves that are wet for as little as three hours at the optimal temperature of 62°F will promote germination of rust spores. The disease has been shown to spread via contaminated cuttings in some instances.

After twenty years with few incidences, geranium rust has recently reappeared on geranium crops.
Photo by A.R. Chase.

The range of geraniums that can be attacked by rust is relatively narrow but appears to include the majority of zonal geranium cultivars. Other species of *Pelargonium* are not as susceptible. Some research conducted in 1979 showed great differences in susceptibility of zonal geraniums. Although most of the current cultivars probably weren't available in the 1970s.

Chemical Control

We conducted a single trial on geranium rust control in 1998. Products were applied as weekly sprays without additional adjuvants added. Compounds tested were: Decree 50W at 32 oz. per 100 gal., Systhane 40WP at 8 oz. per 100 gal., Phyton 27 at 16 oz. per 100 gal., an experimental formulation of mancozeb at 38 oz. per 100 gal., RH-06ll at 18 oz. per 100 gal., and Banner Maxx at 6 oz. per 100 gal.

Excellent control was achieved with Systhane, Banner Maxx, the experimental mancozeb, and the combination of myclobutanil and mancozeb (RH-0611). Phyton 27 gave about 60% control, and Decree gave almost 50% control.

Be careful to read product labels to determine their legal and safe uses on geraniums in your state. Not all of the products mentioned are labeled for rust control on ornamentals.

Cultural Control

As always, in addition to chemical or biological fungicides, use all cultural methods available to reduce rust severity. Never use cuttings that show signs of rust infection. Do not establish your own mother blocks for cuttings for longer than a season without weekly scouting for symptoms and annual block renewal. Rust overwinters and oversummers very effectively on stock plants. Remove diseased plants from the growing area, try to choose resistant cultivars when possible, and increase greenhouse temperatures when feasible to above 62°F. Under these conditions many fungicides give very good to excellent control.

Pest Control, August 1999.

Powdery Mildew versus Downy Mildew

Jean L. Williams-Woodward

Think your crops have powdery mildew? You've been spraying for it, but nothing seems to work? Check again—you may have downy mildew, a disease whose symptoms can mimic its more common cousin.

Powdery mildew is easily recognized by the powdery white to light gray spots on leaf, stem, bud, or flower tissue. Eventually, the infection can cover the entire leaf or stem. Infected tissues can be stunted, chlorotic, curled, or deformed. Succulent, young tissue is most susceptible to infection. Older infections often cause necrosis and premature senescence of leaves. It prefers warm days, cool nights, and dry conditions, unlike many other foliar diseases.

In contrast, downy mildew, a water mold in the same class of fungi as *Pythium* and *Phytophthora,* produces its characteristic symptoms on leaf undersides. Sporulation is seen as fuzzy, white to gray patches. Leaf upper surfaces can have pale green to yellow spots opposite the areas of sporulation. Older leaves are affected first and often turn completely brown from infection. Severe downy mildew infections can cause fuzzy, white sporulation on upper leaf surfaces that can look like powdery mildew infection. It favors high humidity, cool temperatures, and long durations of leaf wetness. Plants severely affected with downy mildew rarely recover.

The same fungicides and cultural controls won't work for both diseases. Control powdery mildew with systemic fungicides including fenarimol (Rubigan), myclobutanil (Systhane), propiconizole (Banner), thiophanate methyl (Cleary's 3336, Fungo), triadimefon (Bayleton, Strike), and triforine (Triforine, Funginex). Protectant fungicides such as chlorothalonil (Daconil 2787), piperalin (Pipron), and trinumizole (Terraguard) also provide good control. Wettable sulfur is also an option.

For downy mildew, fungicide options include metalaxyl (Subdue) and fosetyl-Al (Aliette). Resistance to metalaxyl is a concern, and tank-mixing metalaxyl with mancozeb is an option to reduce resistance potential.

Culture Notes, September 1997.

Poinsettia Fungus Update

Watch for what may be an unfamiliar disease on your poinsettia crops. Growers in New York and several other states found poinsettia scab—a disease normally found on outdoor-produced crops in Florida—on their greenhouse crops last year. Though greenhouses don't usually have conditions conducive to the spread of poinsettia scab, also known as spot *Anthracnose,* you should be ready to manage it if necessary.

The first step is detecting and roguing out infected plants as soon as possible, Margery Daughtrey, plant pathologist, Cornell University, said in Cornell's *Greenhouse IPM Update*. Infected plants won't be salable.

On leaves, poinsettia scab causes small, round, tan spots with purple rims that buckle out from the leaf surfaces. Whitish, scablike (raised) lesions appear on stems and along leaf midribs. Stem symptoms often occur at the base of the plant, where they're less likely to be detected. In severe cases, lesions can coalesce and girdle stems. The most obvious symptom of scab is the unusual elongation of infected stems or branches. Infected plants appear strangely tall compared with the rest of the crop.

Protectant fungicides with active ingredients thiophanate-methyl, mancozeb, and chlorothalonil labeled for controlling poinsettia scab include Cleary's 3336, Domain, Dithane T/O, Protect T/O, Zyban, Benefit, Daconil Ultrex, and Daconil WeatherStik. Use fungicides with cultural practices to minimize splashing and the length of time leaves are wet. Remove infected plants from greenhouses before fungicide applications.

Culture Notes, October 1997.

Doing It Right: Controlling Mildew on Poinsettias

Mary K. Hausbeck, Margery Daughtrey, and Larry W. Barnes

Managing diseases that affect foliage of flower crops requires a thorough understanding of environmental and cultural influences, as well as knowledge of how to monitor for symptoms and how to best utilize the available pesticides. In 1995, a research team consisting of Dr. Mary Hausbeck (Michigan State University), Margery Daughtrey (Cornell University) and Dr. Larry Barnes (Texas A&M University) was formed to address some of these needs for the flower industry.

Initial studies were directed at learning how to manage the new powdery mildew disease on poinsettias, which has been a significant problem for growers in the Midwest and northern U.S. and an occasional problem for growers elsewhere. Losses from this disease have ranged from $10,000 to $140,000 when growers didn't implement effective management plans in time. Powdery mildew may not be detected until late in the production cycle, and by that time bracts are infected and plants aren't marketable. In

some cases, powdery mildew may not develop until the plants leave the production greenhouse, resulting in customer dissatisfaction.

Research Highlights

We've investigated resistance of poinsettia cultivars to powdery mildew, but disease developed on all cultivars tested within one month after inoculation. At 68°F (85% relative humidity), it only takes six hours for a powdery mildew conidium (spore) to germinate on a poinsettia leaf. Within twenty-four hours, this germinated conidium forms a "sac" that the fungus uses to extract nutrients from the plant. However, at 86°F, conidia aren't able to germinate as readily, and their ability to form the sac necessary to thrive is almost eliminated. This helps to explain why epidemics of powdery mildew on poinsettia typically don't occur during the summer in the Midwest: Greenhouse temperatures may frequently exceed 86°F, thereby slowing or stopping disease development.

Currently, we're exploring whether temperature manipulation is a useful tool for managing powdery mildew on poinsettia. High temperature treatments to eradicate disease might be feasible during stock plant production or following rooting of cuttings. It may be helpful to expose rooted cuttings to a high temperature treatment to ensure that they're free of powdery mildew. Though it's unlikely that heat treatments can be used alone, it may be possible to incorporate temperature manipulation with scouting and fungicide applications to reduce growers' powdery mildew management costs.

Fungicide studies are focusing on whether systemic fungicides applied to stock plants provide protection to cuttings removed from those plants. We're also testing currently nonregistered fungicides and biocontrols for their ability to control powdery mildew, to encourage companies to expand labels to include poinsettia.

As a direct result of this American Floral Endowment–sponsored research, we've developed effective management recommendations for powdery mildew on poinsettias. Scouting is the most essential part of a management program. Powdery mildew appears as white colonies on the tops and undersides of foliage and bracts. Unlike *Botrytis,* powdery mildew doesn't require extended leaf wetness periods in order to infect, so growers skilled at environmental control techniques for managing *Botrytis* blight may still encounter problems with powdery mildew. Here are our tips for successfully preventing and controlling powdery mildew on your poinsettias.

Preventive Management Program

Use these tactics when you haven't found powdery mildew in the greenhouse. Scout one out of thirty plants weekly. Examine four fully expanded leaves on the lower and middle portion of each plant (a total of eight leaves), paying special attention to leaf undersides. You could apply these fungicides every fourteen days (table 1), but you should alternate among the active ingredients listed to avoid potential growth regulatory effects of some of the products. We recommend that you apply fungicides before bract coloration to prevent fungicide residue or phytotoxicity.

Curative Management Program

Use these tactics when you've found powdery mildew in the greenhouse. Scout one out of ten plants weekly until plants are free of disease for at least three weeks. After that, resume scouting one out of thirty plants weekly. To assess the effectiveness of the spray program, note whether any detected powdery mildew colonies are living and active. Inactive powdery mildew colonies will be flattened against the plant's surface. Colonies that are alive and active will have a granular, powdery, or fluffy appearance due to the production of conidia (spores).

Table 1. Fungicides for Prevention of Powdery Mildew	
Active ingredient	**Product**
Triadimefon	Strike 25DF (2.0 oz./100 gal.)
Triflumizole	Terraguard 50W (4.0 oz./gal.)
Myclobutanil	Systhane WSP (4 oz./100 gal.)
Thiophanate-methyl	Cleary's 3336-F (20 oz./100 gal.)
	Cleary's 3336WP (24 oz./100 gal.)
	Domain (10 oz./100 gal.)
	Fungo-Flo (10 oz./100 gal.)
	SysTec 1998 (10 oz./100 gal.)
	SysTec 1998WDG (0.4 lb./100 gal.)
Thiophanate-methyl + iprodione	Benefit (17 oz./100 gal.)
Copper sulfate	Phyton-27 (15 oz./100 gal.)

When feasible, remove all leaves infected with powdery mildew; immediately place leaves into a bag and promptly seal it. Don't carry infected leaves through the greenhouse for disposal because spores on the infected tissue will be released and may infect nearby healthy poinsettias. Although removing infected leaves may sound labor intensive, regular scouting should ensure that the number of infected plants per area and the number of infected leaves per plant will be low enough to make leaf removal feasible (one to five leaves per infected plant). Removing leaves at any early stage of disease development disposes of a source of spores that would otherwise contribute to epidemic development during the fall finishing season.

Apply these fungicides every fourteen days (table 2), alternating among the active ingredients listed to avoid the potential growth regulatory effects of some of the products.

Table 2. Fungicides for Curative Management of Powdery Mildew	
Active ingredient	**Product**
Triadimefon	Strike 25DF (4.0 oz./100 gal.)
Triflumizole	Terraguard (16 oz./100 gal. for first application, then 4 to 8 oz./100 gal.)
Myclobutanil	Systhane WSP (4 oz./100 gal.)

November 1997.

Fusarium Wilt of Cyclamen

A. R. Chase

Fusarium wilt, described from Germany in 1930 and in the U.S. in 1949, is one of the most severe diseases affecting cyclamen. The pathogen causing the disease is a species that affects only cyclamen: *F. oxysporum* f. sp. *cyclaminis*. Symptoms include yellowing and wilting of lower leaves, root rot, browning of vascular tissues in the corm, and sometimes corm rot. Infected plants inevitably collapse and die. All ages of plants are susceptible and can develop severe disease. (Early reports that only flowering cyclamen showed symptoms were perhaps due to the seasonal nature of production and effects of temperature on disease development.) Peak losses occur when temperatures reach 80°F or higher. In general, temperature increases shorten the time for symptoms to develop.

The initial source of the pathogen remains debatable. Many researchers believe that *Fusarium* is introduced into a production range via infected or infested seed and remains a problem by living on dead or decaying matter for extended periods, even when cyclamen aren't grown. Seed treatments with hot water or bleach solutions are an integral part of *Fusarium* wilt control for some cyclamen growers.

Cultural Control

All cyclamen cultivars tested to date have been found to be susceptible to this disease. Studies of potting media compositions have indicated that sphagnum peat may be one of the most conducive media components promoting disease development. Both composted hardwood bark and pine bark media have shown some promise for reducing the severity of *Fusarium* wilt in cyclamen. In addition, wilt due to drought is thought to trigger the development of latent infections of *Fusarium* wilt.

One of the most important strategies for control is raising potting media pH above 6.0 by adding dolomitic lime. In addition, using nitrate nitrogen results in lower disease severity than the use of ammoniacal nitrogen. Frequent scouting and removal of symptomatic plants is a critical component of any control program, as therapeutic applications of neither biological nor chemical products have resulted in complete disease control.

Chemical Control

Fusarium wilt has proven extremely difficult to control with any currently available fungicides. In the past, researchers identified benomyl as the most effective fungicide for controlling this disease, but it wasn't very effective. Thus, the researchers recommended the strict sanitation and cultural methods recommended above. However, because new products are periodically introduced, a series of chemical control trials was conducted at Chase Research Gardens Inc. to identify potentially effective products for this serious disease.

All plants were obtained as plugs and planted into new pots (usually four-inch) with peat-based potting media. They were fertilized with Nutricote slow-release fertilizer at a rate of ¼ tsp. per pot and watered as needed. Tests were run throughout a one-year period using various cultivars from the Sierra series. Plants were treated weekly with a drench of the test products at a rate of about 1 pint per sq. ft.

All plants were artificially inoculated with spores of the pathogen added to potting media in a water solution. Symptoms were rated on two scales: The first scale evaluated severity of yellowing, wilting and collapse of plant foliage using the following scale: 1 (no symptoms), 2 (slight symptoms), 3 (moderate symptoms), 4 (severe symptoms), and 5 (dead plant). The second disease rating evaluated corm discoloration by cutting them at the end of the test. The following scale was used: 1 (no browning, white), 2 (slight browning of less than 10% of vascular tissue), 3 (moderate browning of 11 to 50% of vascular tissue), 4 (severe browning of 51 to 90% of vascular tissue), and 5 (complete browning of vascular tissue). Corm browning was typically less severe than foliar symptoms on any given plant.

This test indicated that the best control comes with applications of Terraguard, Medallion, and Heritage. Subsequent tests were performed to evaluate different rates of these three fungicides. Terraguard was tested at 2, 4, and 8 oz. per 100 gal., and good control was obtained with 4 or 8 oz. Medallion was tested at 1, 2, and 4 oz. per 100 gal. In this case, 1 and 2 oz. gave control equal to Terraguard at 8 oz., and Medallion at 4 oz. gave very good control. Additional tests with Heritage showed the best control with a 16-oz. rate, but 8 oz. of Heritage was comparable to 8 oz. of Terraguard or 2 oz. of Medallion.

Despite the fact that several products gave excellent control under conditions of low to moderate disease pressure, they didn't protect some plants from becoming infected and eventually dying of *Fusarium* wilt. The most promising fungicides are Medallion and Heritage. Although several other products affected disease severity, the extreme difficulty in controlling this disease requires a very high degree of control and rules out their reliable use.

Medallion is currently labeled for *Rhizoctonia* disease on some ornamentals and is very effective for a variety of other pathogens. Heritage is available only for use of turf at this time, but the label for ornamental diseases should be written in the next year. Both products, while the most effective available, should be used in conjunction with a complete cultural control program. Only in this setting can they be helpful, as under high disease pressure, especially during the summer months, they won't stop every plant from becoming infected and dying.

Conclusions

These tests indicate potential for using fungicides as a part of an integrated program for controlling *Fusarium* wilt on cyclamen. Used alone, none of the fungicides can be expected to completely control this severe wilt disease. Use pathogen-free seeds or plugs, grow plants in new pots with new potting media, and remove symptomatic plants as soon as you find them. In addition, use nitrate nitrogen, adjust the potting media pH to 6.0 or higher, try to keep temperatures below 80°F, and judiciously apply fungicides.

Fungicide Efficacy for *Fusarium* Wilt of Cyclamen

Fungicide	Manufacturer	Rate per 100 gal.	Degree of control
Chipco 26019 50WP	Rhone-Poulenc Ag Company	16 oz.	None to slight
Heritage 50WG	Zeneca Inc.	4 to 16 oz.	Moderate to excellent (depending on disease pressure)
Medallion 50WP	Novartis Crop Protection Inc.	1 to 4 oz.	Good to excellent (depending on disease pressure)
Phyton 27	Source Technology Biologicals Inc.	25 oz.	None
SP2001 WDG	SePRO Corporation	32 oz.	Poor to moderate
Terraguard 50WP	Uniroyal Chemical Company Inc.	2 to 8 oz.	Moderate to good (depending on disease pressure)

Editor's note: The author thanks Uniroyal Chemical Company Inc., SePRO Corp., Source Technology Biologicals Inc. and Goldsmith Seeds for partial support of these chemical control studies.

Horticultural Oil for Powdery Mildew on Gerbera

David L. Clement, Rondalyn M. Reeser, and Stanton Gill

As part of an ongoing greenhouse TPM/IPM program at the University of Maryland, we conducted trials in 1996–97 to evaluate the effectiveness of several horticultural oils for preventing powdery mildew, *Erysiphe cichoracearum*, on gerbera daisy.

We evaluated three horticultural oils on gerbera daisy cultivars 'Festival Mix' and 'Lemon Yellow' between August 1996 and March 1997. The oils tested were Stylet Oil (JMS Flower Farms Inc., Vero Beach, Florida), SunSpray Ultra-Fine Oil (Sun Co. Inc., Marcus Hook, Pennsylvania), and Triact 90 EC (Thermo Trilogy Corp., Columbia, Maryland).

These products were applied to runoff on four-week-old transplants at rates of 0.5%, 1.0%, and 2.0% (v/v) over a six-to-eight-week period. The standard fungicide control was Cleary's 3336 (W. A. Cleary Chemical Corp., Dayton, New Jersey) applied at the label rate.

Sprays were applied at one- or two-week intervals. Plants were rated weekly during the trials for disease incidence. Temperatures during the spray trials ranged from a low of 60°F at night to 108°F during the day. Relative humidity values ranged from a low of 26% at night to 74% during the day.

Results

Horticultural oil treatments affected flower initiation, foliage color, plant height, and vigor. Plants treated with Cleary's 3336 appeared normal, had little visible residue, and showed no disease when sprayed at weekly intervals throughout the trials. Some plants sprayed with Cleary's 3336 at two-week intervals did develop powdery mildew. Plants treated with Stylet Oil at all spray rates and intervals had consistently smaller leaves, shorter height, and lighter green foliage compared with plants treated with other oils. Plants treated with Triact 90 EC at all spray rates and intervals usually had the largest leaves, greatest height, and darkest foliage. Plants treated with SunSpray Ultra-Fine and Stylet Oil consistently flowered earlier than those in other treatments.

Regardless of the season, applications of SunSpray, Triact 90 EC, or Stylet oil at the 2.0% rate applied either weekly or every two weeks caused a reduction in leaf size, reduced plant height, and prevented flower initiation.

Additional treatment effects for all oils at the 2.0% rate included a shiny residue buildup on the leaf surface, lighter green foliage, and earlier leaf senescence.

We speculate that the 2.0% rate of oil caused tissue damage, reduced respiration rates, slowed the growth rate, and caused additional plant stresses. Oil treatments at the 1.0 or 2.0% rate at temperatures consistently above 90°F caused more problems with plant size, color, and flowering.

The highest rate of oil that could be used without affecting plant size or color was 1.0% every week during temperatures less than 90°F. Applications of SunSpray, Triact 90 EC, or Stylet Oil at the 1.0% rate every two weeks gave few problems with phytotoxicity symptoms and still gave adequate disease control.

All three oils applied at the 0.5% rate at two-week intervals gave the fewest phytotoxicity symptoms, but disease control was inadequate in some cases. Triact 90 EC applied at the 0.5% rate at two-week intervals gave good control of powdery mildew. The 0.5% rate of SunSpray Ultra-Fine Oil or Stylet Oil at two-week intervals didn't consistently control powdery mildew.

Based on our trials, the grower incorporated the use of 1.0% horticultural oil rates of SunSpray or Triact 90 EC at one- or two-week spray intervals depending on the severity of disease throughout the remaining 1997 growing season with good success.

Culture Notes, April 1998.

◪

Downy Mildew: Controlling This Serious Problem

A. R. Chase

Downy mildew diseases are currently causing serious losses in many floriculture crops, including roses, snapdragons (both cut and bedding types), and pansies. And new hosts of downy mildew are being found each year—most recently alyssum, salvia, and rosemary.

Downy mildew of roses is caused by *Peronospora sparsa* and was first found in England in 1862. It's erratic, appearing in some years but not in others. Downy mildew on snapdragons was described in 1936 from Ireland. It's caused by *Peronospora antirrhini*. Although there are a number of downy mildew diseases reported for floricultural crops (table 1), the only two that have been studied significantly are the downy mildews of snapdragons and roses. The diseases on each plant type can differ, and key temperatures for development and other environmental parameters remain unknown for the majority of downy mildew diseases on many bedding plants, perennials, and cut flowers.

Table 1. Some Downy Mildew Pathogens on Floricultural Crops	
Plant	**Pathogen**
Anemone	*Plasmopara pygmaea*
Aster, china aster, goldenrod, and erigeron	*Basidiophora entospora*
Carnation	*Peronospora dianthicola*
Centaurea, *Crigeron*, goldenrod, rudbeckia, and senecio	*Plasmopara halstedii*
Clarkia, *Gaura, Godetia*, and evening primrose	*Peronospora arthuri*
Cornflower and other composites	*Bremia lactucae*
Geranium	*Plasmopara geranii*
Lupine	*Peronospora trifoliorum*
Pansy, viola, and African violet	*Bremiella megasperma* or *Peronospora violae*
Rose	*Peronospora sparsa*
Snapdragon	*Peronospora antirrhini*
Veronica	*Peronospora grisea*

Some downy mildew diseases are more aggressive than others are. For example, downy mildew on snapdragons appears to spread much faster and cause more serious losses faster than downy mildew on pansy and viola. Because the fungus grows within the plant tissues and not on the surface, it can escape notice until the conditions are ideal for sporulation.

At this time, the fruiting structures of the fungus emerge from the undersides of leaves and create the grayish colored, downy coating. On some plants, this may be the first indication that they're infected with a downy mildew fungus. In other plants, distortion of new leaves, downward curling, and overall stunting occur, mimicking aphid damage. Other plants, such as roses, develop reddish black spots on leaves, petals, and stems, well in advance of sporulation.

Where Does Downy Mildew Come From?

Downy mildew control should start with use of clean, pathogen-free seed, as some downy mildews, such as sunflower downy mildew, are transmitted through infected seed. Some downy mildew diseases of ornamentals are believed to be seed transmitted, but this has yet to be proven. In addition, many weed hosts of downy mildew fungi attack cultivated crops. It's possible that some epidemics begin from infections on native plants and weeds in and around production areas.

Rose downy mildew sometimes starts on bare-rooted, apparently healthy stock. Rose canes infected with the pathogen may not be obvious, and disease may crop up only when environmental conditions are ideal. Exposing spores to 80°F for twenty-four hours kills them. Thus a heat treatment of canes, seeds, or other propagative stock might be possible to kill spores on their surfaces. Killing the pathogen within the plant would be more difficult, and the high temperatures needed might damage the plant along with the downy mildew.

Humidity and Irrigation Management

Downy mildew is most severe when nights are cool and days are warm with high relative humidity. Humidity management is sometimes possible and always desirable when growing plants in a greenhouse. It's critical to keep the relative humidity below 85% to decrease sporulation on infected plants and stop germination of spores on new plants. This can be done in greenhouses by venting and raising the temperature at key times during the day. The best time for this is at sunset, when the greenhouse air is warm and moisture-laden and the outside air is cool and dry. Venting followed by heating will fill the greenhouse with warmer, drier air. Improving drying of wet leaves with fans is also recommended but can cause problems with disease spread, as downy mildew spores are easily spread by air movement. Use other methods (fungicides or removing infected plants) with fans to keep spores from spreading this way.

Early work on snapdragon downy mildew showed that exposure to rainfall wasn't necessary for disease development, but high humidities (above 85%) were critical. Nevertheless, keeping the leaves dry will keep spores from germinating at relative humidities below 85%.

Temperature Effects

The optimal temperature for the development of rose downy mildew is 64°F, and snapdragon downy mildew develops best at temperatures between 40° and 60°F. Optimum temperatures for other ornamental downy mildew fungi aren't known at this time. A few of those known for non-ornamental crops include crucifers such as cabbage and broccoli (45° to 60°F), lettuce (50° to 70°F), and soybeans (50° to 80°F). Thus, although the temperatures are close, they aren't identical, and each disease must be studied to determine the optimal range for that downy mildew fungus.

Sanitation

Sanitation requirements for downy mildew are stringent. Infected plant tissues such as leaves, stems, and flowers may drop to the ground, where the spores can remain viable for a very long time. On some crops (sunflower), infections occur through germination of oospores found in the soil from the previous year's crop.

Remove all infected plants, and discard them well away from your production area. If you collect debris in a pile close to production, you may continue to experience new infections starting with the formation of spores on plants in the trash pile.

Fungicides

The most important thing about using a fungicide to control downy mildew is to recognize the relationship between these fungi and other plant pathogens. The downy mildews are only distantly related to powdery mildew. They are related to *Pythium* and *Phytophthora*. Thus, fungicides that are effective for pythiaceous fungi have the best activity against downy mildew fungi.

Fungicides have been tested for control of downy mildew on roses throughout the world. In general, dithiocarbamates (such as Dithane T&O and Protect), sulfur dusts, and, occasionally, copper products have been recommended. Although metalaxyl (Subdue) provides excellent control when used alone, resistance can develop, making the addition of a dithio-carbamate necessary to maintain good disease control. (Novartis marketed Pace for this use in the U.S.) Fosetyl aluminum (Aliette) was reported as

---

Table 2. Chemical Control of Downy Mildew

Plant	Fungicide spray	Rate per/100 gal.	Result
Pansy	Subdue Maxx	0.5 or 1 oz.	Complete control at both rates
	Protect T&O	16 oz.	Complete control when used every other week or every week
	Heritage	1, 2, or 4 oz.	Complete control at all rates
Snapdragon	Phyton 27	15, 30, or 45 oz.	Slight disease at 15 oz., complete control at 45 oz. (some foliar burning)
	Aliette 80WDG	2 or 4 lbs.	Complete control at 2 lbs. (some damage at 4 lbs.)
	Bavol	15 oz.	About 50 % control

This table represents results of two tests on chemical control of downy mildew on snapdragon and pansy. The work was funded by the following companies: Chase Research Gardens, W.A. Cleary, Source Technology Biologicals, and Zeneca.

effective when used at 3 lbs. per 100 gal. at fourteen-day intervals. Finally, propamocarb was tested in Australia and found to give excellent control as a soil drench when used every four weeks. For nonsystemic products, it's critical to make applications to leaf undersides, as sporulation on many plants occurs there.

The results of two recent experiments on bedding plants infected with downy mildews are given in table 2. All products were applied as foliar sprays on a weekly basis. These tests indicate a large number of possibilities for control, including systemic products such as Subdue and protectants such as Protect T&O. Be sure to read pesticide labels carefully for legal uses on your crops under your conditions.

October 1998.

Fusarium Diseases Targeted by Multistate Research Team

Robert J. McGovern, Wade H. Elmer, David M. Geiser, and Brent K. Harbaugh

Some of our industry's most serious disease problems are caused by different species of the fungus *Fusarium*. In some *Fusarium* diseases, only the roots of plants are affected, while others feature cankers at the stem base.

Sponsored by the American Floral Endowment, a team of researchers from the University of Florida, the Connecticut Agricultural Experiment Station, and Pennsylvania State University are working to develop new, practical knowledge about this notorious group of pathogens. This long-term, interdisciplinary project will study the genetics, biology, epidemiology, and integrated management of diseases of potted ornamentals caused by *Fusarium*. The specific diseases and crops we'll study are *Fusarium oxysporum* on cyclamen; *F. solani* on caladium; *F. oxysporum* and *F. solani* on chrysanthemum; and *F. avenaceum*, *F. solani*, and *F. oxysporum* on *Lisianthius*. Much of this information will be applicable to the control of *Fusarium* in bedding plants, cut flowers, and other potted ornamentals.

The Problems

Diseases caused by *Fusarium* species have been reported in most major potted ornamentals. Severe outbreaks of *Fusarium* crown and stem rot caused by *F. avenaceum* have greatly decreased production of *Lisianthius* in the U.S. in recent years. *Lisianthius* production is also seriously threatened by two other poorly understood diseases incited by *Fusarium*: a root and crown rot (*Nectria haematococca*) caused by *F. solani* and a vascular wilt caused by *F. oxysporum*.

Florida studies have recently shown that the disastrous decline in caladium tuber production over the past decade is strongly linked to *F. solani*.

Vascular wilts caused by *F. oxysporum* represent continuing challenges for the production of chrysanthemum and cyclamen. Considerable losses to chrysanthemum have also been caused by *F. solani*.

Very little new information has been developed on *Fusarium* diseases in the past twenty years. A unified, in-depth approach that focuses on several *Fusarium* species and plant production systems is needed to develop effective management techniques. The team's objective is to generate useful information on the genetics, biology, epidemiology, and integrated management of the wide range of *Fusarium* diseases.

Research Approaches

Genetic characterization

Recent molecular research methods have indicated that ornamentals may be infected by an even greater diversity of *Fusarium* species than is currently recognized. Studies hold promise for disclosing new information about host range and geographic origin of the various *Fusarium* fungi.

Molecular techniques have also been useful in studying fungicide resistance. In this study, we'll look for subgroups within *Fusarium* species associated with fungicide resistance and use molecular methods to develop specific detection techniques that will be of immediate benefit to plant propagation programs.

Biological and epidemiological studies

Knowing how pathogens survive and spread and the conditions that favor infection is essential for effective disease management. There are major data gaps in the biology and epidemiology of *Fusarium*-incited diseases in each of the four targeted crops, especially caladium and *Lisianthius*. For instance, we need to focus on the role of fungus gnats in the spread of *Fusarium* diseases. Studies thus far have shown the importance of fungus gnat spread during the production of *Lisianthius* seedlings.

Does *Fusarium* spread through the air? Although it's not generally considered to be air-disseminated, we've recently obtained spore trap data suggesting that *F. avenaceum* may be spread to a limited extent by air to *Lisianthius*.

Promising New Cyclamen Research

At the University of Florida, fungicides were evaluated for reduction of *Fusarium solani* in caladium in conjunction with a hot-water treatment (122°F for 30 minutes). The incidence of infection was rated by placing tuber cores on selective media. Banrot, Armicarb 300, Consyst, and Heritage treatments significantly reduced *Fusarium* incidence.

At the Connecticut Agricultural Experiment Station, researchers examined the effect of salt (NaCl) for suppression of *Fusarium* diseases of plants. One prerequisite for this strategy to work is that the host plant must be tolerant to salt. Unfortunately, most ornamentals lack this trait—but cyclamen is an exception!

Experiments were conducted with cyclamen seedlings grown in soil infested with *F. oxysporum* and treated with increasing rates of NaCl. Seedlings were also grown in noninfested soil and treated the same way. At the end of the trial, most infected plants grown without NaCl died, but plants treated with NaCl between 0.25 and 0.5% were alive and larger. Rates higher than 1% tended to cancel the disease-suppressive benefits and caused salt stress.

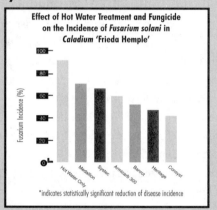

Effect of Hot Water Treatment and Fungicide on the Incidence of *Fusarium solani* in *Caladium* 'Frieda Hemple'

*indicates statistically significant reduction of disease incidence

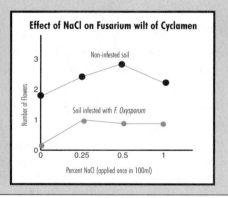

Effect of NaCl on Fusarium wilt of Cyclamen

We'll also investigate other ways *Fusarium* may be spread. For instance, the role played by infected seeds in *Fusarium* wilt of cyclamen has long been suspected but never confirmed. And vegetative spread is another important aspect: Infected chrysanthemum cuttings and caladium seed tubers are undoubtedly important in the perpetuation of *Fusarium* wilt and tuber rot. We'll also study the survival of pathogenic *Fusarium* species on greenhouse surfaces, in soil, in irrigation water, and on alternative weed and crop hosts.

Integrated management strategies

Exclusive reliance on chemical control may be risky. Resistance, especially to the benzimidazole class of chemicals, has been documented in a number of

Fusarium species. Our goal is to develop an integrated approach to *Fusarium* management by identifying cultivar resistance, inducing resistance, biological control, chemical control, and cultural control using plant nutrients and by manipulating pH. Chemical control will focus on low-impact fungicides with short reentry intervals.

What the Future Holds

We expect that the fundamental and practical knowledge of *Fusarium* we'll gain from this project will be applicable to a wide range of ornamentals. The benefits to the industry will include more precise, safe, and effective methods for diagnosing and managing *Fusarium* diseases.

January 2000.

Knock out Root and Crown Rot with Fumigants or Biologicals

R. Kenneth Horst

Verticillium and *Fusarium* fungi cause destructive plant diseases and both are soil inhabitants. Other species that cause root rots and sometimes crown rots are *Pythium, Rhizoctonia, Sclerotinia, Sclerotium, Thielaviopsis,* and *Phytophthora. Pythium, Rhizoctonia, Sclerotinia, Fusarium,* and *Phytophthora* all cause root and/or rot of Easter lilies grown as an early spring pot crop. Crop rotation is used to control some diseases caused by soilborne fungi. This disease control procedure is not practical for controlling *Verticillium* because the fungus can attack such a wide range of plant species, and the pathogen persists as microsclerotia in soil. Microsclerotia are resting structures of the fungus that survive in the absence of known hosts for many years.

The Fumigant Solution

Soil fumigants are commonly used to reduce disease severity in fields that contain these pathogens.

Soil fumigation procedures reduce the amount of pathogens in the soil and thus decrease disease severity or the onset of symptoms in fields containing pathogens. Control of *Verticillium* and *Fusarium* is only partially successful with the application of fumigants. Disadvantages of fumigation include the adverse effect of chemicals in the environment and the high costs of applications. Moreover, the availability of many fumigants is questionable due to their adverse environmental impact.

Many root pathogens are difficult to control. A number of fungicides are registered and sold for greenhouse soil use, but results of fungicide treatment may vary depending on such soil conditions as texture, moisture, and temperature, which require dosage adjustment. Heavy, dry, or cold soils generally require more fungicide to get acceptable results. Also, selecting most soil drench fungicides has to be based on their effectiveness in controlling specific pathogens.

The Biological Alternative

In recent years, greenhouse production practices have changed, as dictated by production costs, consumer trends, and availability of new products. Currently, concerns over worker protection, reentry intervals, and environmental protection have forced adjustments in approaches to pest and pathogen control. The impending loss of the widely used fumigant methyl bromide has given impetus to researching and developing alternative methods of controlling soilborne pests and pathogens.

Currently, there are four microbial biopesticide active ingredients for use on ornamentals and turf registered by the U.S. Environmental Protection Agency. Bio-Trek and RootShield *Trichoderma harzianum* KRL-AG2 are formulations of a selected strain of a common soil saprophytic fungus. Incorporated into growing media or used as a seed treatment, they protect against *Pythium, Rhizoctonia solani*, and *Fusarium* on greenhouse ornamentals.

Galltrol A and Norbac 84C *Agrobacterium radiobacter* K84 are bacterial cultures used as a root dip, primarily on woody ornamentals to prevent crown gall.

Mycostop *Streptomyces griseoviridis* K61 is a powdered formulation of dried spores and mycelia of the soil actinomycete. It can be used as a seed treatment, transplant or cutting dip, soil spray or drench, or through drip irrigation systems. It's labeled for use against seed rot, root and stem rot, and wilt caused by *Fusarium, Alternaria, Phomopsis, Pythium, Phytophthora,* and *Rhizoctonia.*

SoilGard 12G *Gliocladium virens* G1-21 is a granular formulation of a common soil saprophytic fungus. Incorporated as a preplant soil amendment, it protects against the soilborne fungal pathogens *Rhizoctonia solani* and *Pythium,* which cause damping-off and root rots of ornamental and food crop plants grown in greenhouses, nurseries, and interiorscapes.

Pest Control, December 1998.

Managing Powdery Mildew on Gerberas

Larry W. Barnes, Mary K. Hausbeck, and Margery Daughtrey

Of the production-related problems that affect gerbera daisies, powdery mildew is probably the most consistently troublesome. Resulting from infection by the fungus *Erysiphe cichoracearum,* epidemics of powdery mildew can seemingly develop overnight, resulting in significant quality loss if you don't achieve disease control.

As part of American Floral Endowment–supported research, we've looked at ways to more effectively and efficiently manage powdery mildew in various floral crops. Our research has demonstrated that good production practices and the use of effective fungicides can properly manage gerbera powdery mildew.

Diagnosis Isn't Difficult

The obvious white, powdery fungal growth on the surface of plant tissue makes powdery mildew diagnosis easy. Although powdery mildew is primarily a leaf pathogen, all aboveground plant parts can become infected, including the flower stalk and flower petals.

The telltale sign of early infection almost always occurs on the upper leaf surface during the initial stages of infection and disease development and usually starts out as a small, white, fuzzy spot. On occasion, powdery mildew may also produce the powdery growth on the undersurface of a leaf, with a small, subtle, but noticeable yellow blotch on the upper leaf surface immediately opposite the powdery stage on the leaf's undersurface. Either of these signs or symptoms indicates the early stages of powdery mildew development and should mark the action stage for management. Fungicides can be highly potent tools for powdery mildew management.

Relative Humidity

Growers should be aware of the influence that temperature and relative humidity have on the development of powdery mildew in gerbera daisy. Moderate temperatures (55° to 85°F) favor powdery mildew, but changes in greenhouse relative humidity that result from normal day-night temperature differences are also important factors in gerbera powdery mildew development.

The inverse relationship between temperature and relative humidity means that as the greenhouse temperature falls at night, the greenhouse humidity increases. High humidity stimulates powdery mildew spore germination, resulting in additional leaf spots, and also encourages additional spore production in existing leaf spots.

As the temperature warms after sunrise, the relative humidity falls, drying the white mass of powdery mildew spores on the leaf surface. As the spore mass dries, it "fluffs up," becoming very powdery, and the spores are readily blown from the leaf surface and spread throughout the greenhouse by air currents. The extremes in relative humidity, caused by changes in temperature, affect both the production and spread of powdery mildew spores.

AFE-Supported Trials

Greenhouse trials were conducted December through April of 1996–97 and 1997–98 to evaluate several fungicides for their effectiveness for control of powdery mildew in gerbera daisy production.

In addition, two other chemicals, soluble silicate (potassium metasilicate from Colorado Hydrogardens) and hydrogen peroxide, were experimentally evaluated. Note: Neither of these two materials is specifically labeled for powdery mildew control, but research has shown silicate products to be effective in managing powdery mildew and some other diseases in other crops; many growers had asked about the efficacy of hydrogen peroxide.

We potted gerbera daisy plugs into four-inch pots and placed them in the greenhouse. A dozen mature six-inch gerberas with extensive powdery mildew infection were randomly placed throughout the greenhouse to provide powdery mildew inoculum for the test plants.

We know that IPM strategies can reduce fungicide usage and production costs if used in conjunction with a thorough scouting program. We wanted to evaluate the efficacy of the fungicides when used in a manner to support a scouting program, so our production area was scouted daily and fungicides weren't applied until the scouting report indicated the first evidence of powdery mildew on the test plants. All fungicides were then applied on a schedule according to label instructions; the soluble silicate and hydrogen peroxide materials were applied at seven- to ten-day intervals.

Fungicide Test Results

Fungicides were evaluated for effectiveness of powdery mildew control and evidence of phytotoxicity in the 1996–97 trial.

Both Banner and Pipron provided complete control of powdery mildew. No difference in powdery mildew control was noted between the 4-oz. and

8-oz. Pipron treatments, either with or without a surfactant. Plants in the Banner and Pipron treatments appeared to be slightly more compact and somewhat darker green than other plants. This growth regulator effect was not judged objectionable and would likely, in fact, be beneficial in many production situations.

How the Fungicides Fared		
Fungicide	**Control**	**Comments**
Banner	Excellent	PGR effect: not objectionable.
Pipron	Excellent	Seven- to fourteen-day intervals needed under high PM pressure.
Terraguard	Excellent	PGR effect: not objectionable.
Systhane	Excellent	PGR effect: not objectionable.
Triact	Excellent	Use at seven-day intervals when PM present; 0.5% & 1.0% rates equally effective.
ZeroTol	Very good to excellent	Use at seven-day intervals.
Strike	Good to very good	2 oz. rate needed for acceptable control.
Soluble silicate	Good	Noticeable residue; might be a reasonable product for PM management in cut gerberas where minor leaf infection and chemical residue may not be objectionable.

Triact treatments also provided excellent powdery mildew control. Although a small percentage of plants did show infection, the number of lesions per plant was small and the lesions didn't appear to increase on the infected plants.

The initial hydrogen peroxide treatment of 2.0% caused significant phytotoxicity from which the plants failed to totally recover. Subsequent 1% hydrogen peroxide applications appeared to provide powdery mildew control compared to the control treatment, but plant injury from the 2.0% application made a fair treatment evaluation difficult. The results, however, were encouraging enough to warrant reevaluation in subsequent trials.

Soluble silicate treatments at 500 and 1,000 ppm provided good powdery mildew control. Although a relatively high percentage of the leaves showed some infection, the treatments inhibited powdery mildew lesion-size increase and appeared to significantly reduce powdery mildew sporulation. This treatment might be more useful for cut gerbera production, where good powdery mildew control is needed but where cosmetic appearance of the foliage is of minimal importance.

We didn't see a noticeable effect on width and height of the plants in any treatments except in the untreated control and the Triact treatments. The massive powdery mildew infection in the controls simply reduced plant vigor, resulting in reduced overall plant growth. Plants in the Triact treatments were larger than plants in all other treatments because the Triact didn't seem to have any growth regulator effect.

Of the systemic fungicides tested, both Terraguard treatments and both Systhane treatments provided total powdery mildew control. Strike was less effective but did result in significant improvement over the nontreated control. No objectionable phytotoxic effect was noted, but all of the systemic fungicides did result in some apparent growth regulator effect, characterized by somewhat darker green, more compact growth habit.

Triact applied at 0.5% at seven-day intervals also provided 100% powdery mildew control. When the retreatment interval was lengthened to fourteen days, we lost some of the powdery mildew control.

ZeroTol was included in the 1997–98 trial as a commercially available replacement for the hydrogen peroxide treatment of the 1996–97 trial. A 1.0% application rate at seven-day intervals provided very good powdery mildew control with no noticeable phytotoxicity.

Soluble silicate treatments also provided control of powdery mildew when compared to the nontreated control. Both powdery mildew control as well as leaf residue increased as the treatment concentration increased from 1,000 to 2,000 ppm.

Prevention through Scouting

A combination of careful scouting and timely fungicide application can provide excellent control of powdery mildew in gerbera daisy. Scouting should begin when the gerbera plugs are unpacked and should continue at two- to three-day intervals throughout the production of the crop. Although powdery mildew infection usually occurs on the upper leaf surface, be sure to examine the undersides of any leaves that have small spots or patches of yellow visible on the upper leaf surface.

Scouting Tips

1. Scout early and often. Start when plugs arrive and continue at two- to three-day intervals.

2. Rogue powdery mildew-infected plants or plant parts detected during roguing.

3. Flag any location where powdery mildew has been found, and scout that general area daily for awhile to make sure powdery mildew control is being achieved.

4. Use an effective fungicide for PM control, being sure to achieve thorough spray coverage.

Remove infected leaves or plants found during scouting if the incidence is low. Carry a plastic garbage bag and carefully transfer infected plants to the bag to minimize powdery mildew spore dissemination. If possible, scout and handle plants during the morning when spores are less readily spread. Flag or otherwise identify any area where powdery mildew was detected during scouting, and revisit that area on a daily basis to determine if powdery mildew spread is occurring and whether disease control efforts are working acceptably.

Once powdery mildew has been confirmed by scouting, be sure to start a fungicide spray program. Continue to scout the production area to make sure that the disease control is being achieved and to determine whether the spray schedule should be modified or the fungicide changed.

July 1999.

Extinguishing Tulip Fire

Carol Puckett

In the Pacific Northwest, ornamental bulbs are big business. What makes this area ideal for bulb production is the temperate climate and abundant rainfall. However, these conditions can be perfect for producing disease as well.

Researchers from Washington State University at Puyallup, including Gary A. Chastagner, have devoted more than twenty years to testing fungicides to determine their effectiveness on field-grown ornamental bulb crops. They've found that dicarboximide fungicides effectively control a broad spectrum of diseases of field-grown ornamentals, without phytotoxicity concerns. In fact, data collected from these field trials at the WSU's Research and Extension Center show that iprodione and vinclozolin provide better control of *Botrytis* blight on tulips than benomyl, anilazine, and dithiocarbamate fungicides.

Botrytis on Tulips

Tulips rank high on the list of important bulb crops in the Pacific Northwest, second only to daffodils. One disease in particular, *Botrytis* blight (*Botrytis tulipae*), also known as tulip fire, can be devastating if left untreated. In addition, *Botrytis* causes considerable postharvest losses during storage and shipment.

There are two main types of symptoms observed with *Botrytis* blight: nonaggressive or aggressive leaf spots. Nonaggressive symptoms appear as small, water-soaked spots that turn tan or brown. These spots may not kill the plant, but they will decrease the quality and lessen the value of the tulip crop. The aggressive leaf spots are tan or whitish-brown with water-soaked margins. These expand and produce additional spores that spread the disease. The disease can also spread when infected petals drop and come in contact with other leaves. These pathogens can cause extensive damage in a short time.

Moreover, the *Botrytis* pathogen overwinters as sclerotia in the soil and on infected bulbs. Shoots that are infected before emergence produce "fire heads" (large masses of spores) on emerging shoots. These spores spread the disease to other leaves. The loss of leaves leads to a severe reduction in bulb yields.

Iprodione (Chipco 26019) was the only chemical from the trial that provided significant control of tulip fire on flowers under high disease pressure. Findings show that neither varying the volume of water used to apply iprodione nor increasing the interval between applications (from fourteen to twenty-one days) had any adverse affects on disease control. None of the other fungicides tested significantly reduced disease levels when disease pressure was high. Research also showed that such applications significantly increase bulb yield.

Cultural Controls

In addition to chemical control, cultural practices help control tulip fire. Here are some tips to ensure plants are protected:

- Make sure planting stock is healthy and inoculum free.
- Dispose of infected fire heads to prevent the spread of spores.
- Remove flowers prior to petal fall.
- Harvest and rotate crops annually to break the disease cycle.

Culture Notes, May 2000.

Viruses

A New Weapon to Fight INSV and TSWV

Karen Robb, Christine Casey, Anna Whitfield, Leslie Campbell, and Diane Ullman

In spite of tremendous advances in crop production technology and integrated pest management, essentially no effective methods are available to help growers control epidemics of tospoviruses such as impatiens necrotic spot virus (INSV) and tomato spotted wilt virus (TSWV). These viruses are transmitted by several species of thrips, among which the western flower thrips is considered the most important. Other thrips found in the greenhouse, such as the common flower thrips and the greenhouse thrips, don't transmit tospoviruses.

Combating tospovirus epidemics is difficult because the thrips vectors are abundant, have many plant hosts, and are frequently resistant to available insecticides. These problems are compounded because the viruses also have a large host range that includes ornamentals, vegetables such as tomatoes, and many weeds. Indeed, virtually all important bedding plant, flower, and pot crops—with the exceptions of roses, poinsettias, and zonal geraniums—can be damaged by these viruses.

At the University of California, Davis, we've developed a new monitoring system that combines yellow sticky traps and tospovirus indicator plants to detect thrips that can infect plants.

Growers need to understand how thrips develop and transmit tospoviruses to successfully use the information from monitoring to control tospoviruses. Perhaps the most critical point to understand is that individual thrips can infect a plant only if they acquired the virus as an immature insect. Infective adult thrips can transmit the virus to healthy plants by feeding for as little as fifteen minutes and retain the ability to transmit the virus throughout their adult lives. A thrips that didn't feed on an infected plant while immature cannot acquire or transmit the virus as an adult, even if it feeds on infected plants as an adult. Immature thrips don't have wings and, if undisturbed, don't generally move off the plant on which they were born until they pupate. As a consequence, only those plants that host the virus and support thrips reproduction produce infectious thrips and are important to virus spread.

The challenge in developing a monitoring system is to devise a strategy to find these plants so they can be targeted for removal or for thrips control when appropriate. Those plants that are virus hosts but don't support thrips reproduction are considered dead-end hosts for tospoviruses because they don't produce infective thrips and, hence, they don't contribute to continued virus spread. Directing management strategies at dead-end hosts won't help reduce the spread of the virus.

Early detection of plants that are producing infective thrips and assessment of thrips numbers is essential to successfully suppress virus spread.

Monitoring for thrips and tospoviruses has traditionally consisted of using sticky traps and plant samples to detect thrips and visual assessments of plants to detect virus symptoms. The laboratory test used to detect tospovirus infection is called an ELISA (enzyme-linked immunosorbent assay). This test is based on a reaction between the virus and specific antibodies.

As part of our research, a new method for detecting TSWV in plants is under development. Called a tissue blot immunoassay, it's similar to the ELISA method of diagnosis in that antibodies to the virus are also used for virus detection. In this test, suspicious plant tissue is cut with a razor blade and pressed onto a special type of paper called a nitrocellulose membrane. This paper then is treated with a sequence of solutions. At the end of the assay, infected plants leave a distinctive purple mark, and healthy plants leave either no mark or a green mark where plant sap stained the membrane.

The current test is specific for TSWV, but future plans include adapting it to detect several tospoviruses simultaneously. Consistent with our goal of producing technology that growers can use, the tissue blot immunoassay is easy to perform because no plant grinding or smashing is necessary and virtually no special equipment is needed. The assay also has potential to be more portable than ELISA and to provide faster results. We expect that as this test is perfected for grower use, it will play a critical role in monitoring.

In our effort to develop a monitoring system, we've used petunia indicator plants as a way to locate sources of infective thrips. These plants show distinctive local lesions when infective thrips feed on them. As shown in figure 1a, lesions appear as small, brown-to-black spots on the leaves (on left) and look very different than the whitish feeding scars left by noninfective thrips (on right). Local lesions result from a hypersensitive response that is the strategy the petunia uses as protection from the virus. In a hypersensitive response, the tissue around the virus entry site dies rapidly, preventing the virus from spreading and causing a system-wide infection in the plant. Local lesions are apparent on petunias about three to seven days after feeding by an infective thrips. Figures 1b and 1c show the early (1b) and late (1c) stages of lesion development. Note how the dead tissue in the center of the lesion shown in figure 1c has turned tan and is surrounded by a dark border. If indicator plants are used routinely at standard locations inside and outside production areas, they can provide growers with invaluable information about where infective thrips are located or where they enter a production area. Control efforts, whether they include pesticides, exclusion strategies or removal of weeds, can then be directed to those areas where they will do the most good.

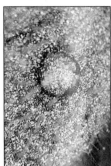

Figure 1. A) Petunia lesions caused by the feeding of infective thrips can be seen on the left leaf, while the whitish feeding scars caused by noninfective thrips can be seen on the right leaf. B) A close look at a lesion as it first appears on the petunia leaf. C) An older lesion on a petunia leaf. As the lesion ages, its center changes from black to tan.

Petunia indicator plants also give growers the advantage of knowing when infective thrips are on the move in an area, even if the crop isn't yet showing symptoms. This is important because many tospovirus-sensitive plants, such as chrysanthemums, may be infected at any time in production, but symptoms aren't visible until the plant sets buds. Relying on symptoms to indicate virus presence or on feeding damage to indicate thrips presence generally doesn't allow the grower to respond soon enough to limit virus spread. The rapid appearance of local lesions on petunias allows for a timely response in deploying thrips control strategies. Although lesions may not immediately be obvious to the untrained eye, growers and scouts can easily learn to recognize them.

Our research has focused on using selected petunia cultivars as indicators of tospovirus transmission by thrips. These are 'Blue Carpet', 'Cascade Blue', 'Summer Madness', 'Burgundy Madness', 'Red Cloud', and 'Super Magic Coral'. Other plants such as fava beans have been evaluated as indicator plants, but the most reliable cultivar of fava beans, 'Toto', is no longer available. Petunias are an excellent choice as an indicator because plants don't support thrips development and seldom become systemically infected. As a result, the plants don't serve as a source of the virus or additional thrips.

We've demonstrated the efficacy of monitoring for infective thrips using petunia indicator plants in conjunction with directional sticky traps. In trials conducted in field-grown flowers, monitoring stations were placed in the field and at the edges of the field. Each station contained directional sticky traps (north-, south-, east- and west-facing traps) and a plant stand for the petunias. It's important that the petunias be at or slightly above the crop canopy and that they be placed on a blue surface to increase their attractiveness to the thrips. Plants are placed in self-watering containers so they don't dry out in the field. Sticky traps and plants showing lesions must be replaced once a week. A sample monitoring station is shown in figure 2. In our studies, we made observations in the field and removed all petunias weekly, held them in the laboratory for a few days, and examined them again for lesions.

For this monitoring system to work, it's essential to grow the petunias in an area isolated from thrips and tospovirus-sensitive plants. Otherwise, growers won't know whether the lesions they observe on the petunia indicator originated in the petunia propagation area or in the production area being monitored. Plants can be used while they're still relatively small (3½-in. pot). Flowers should be removed from plants before placing them at the monitoring station. This is important because the thrips are more attracted to flowers than foliage, and petals don't express local lesions.

Figure 2. A typical trapping station with directional sticky traps and a petunia plant on a blue background.

The information gained from using the four directional traps at each monitoring station has provided valuable insight about the direction from which thrips enter the field. In our trials, the greatest numbers of thrips were consistently caught on the north-facing sticky traps, and the first lesions were detected on petunia at trapping stations at the north end of the field. This result directed our attention to the fields north of the production area where a large block of TSWV-infected, thrips-infested *Malva* (a noxious weed) was discovered.

The grower quickly focused on removing the *Malva* from his field and from the surrounding areas to the north. Directing this control effort to a specific area made it feasible, and the result was a dramatic decrease in spread of TSWV to the grower's flower production area.

"Why use indicator plants at all?" and "Why not just spray the crop regularly or when thrips are found on sticky traps?" Growers often ask these questions before trying our monitoring system. Although many growers have tried routine spraying, they often find that they still have problems with INSV or TSWV. Sticky trap counts alone don't necessarily reflect the number of infective thrips present, nor do they reveal their source.

There is no relationship between the average number of western flower thrips collected on sticky traps and the average number of lesions found on petunias. This is because only the infective thrips in the population can cause lesions on the petunia, and these are the only thrips important to virus spread. Because one insect can infect several plants, it isn't surprising that low levels of infective thrips can reflect a high level of virus. In our trial, peak lesion numbers occurred in an area where western flower thrips populations were relatively low. Conversely, we observed peak numbers of western flower thrips on sticky traps where we observed relatively few lesions.

Usage of the petunia indicator plant/directional trap system alerts growers to the presence of infective thrips and helps locate their source. In our experience with the system, removing these sources resulted in greatly reduced virus incidence. For example, in our trials with field-grown flowers, the number of infected plants dropped from 70% to less than 1% in the first year the monitoring system was tested. We're currently expanding our research to determine the optimal number of trapping stations and the best strategies for indicator plant placement in different types of greenhouses and crops.

Note: This research has been generously supported by the American Floral Endowment, the Carlsbad Agricultural Improvement Fund, Mellano and Company, the University of California Integrated Pest Management Project, and the Binational Agricultural Research and Development Fund. We're grateful to Dr. John Sherwood, University of Georgia, for providing antibodies to the nonstructural protein encoded by the S RNA of TSWV for our use in the tissue blot immunoassay.

February 1998.

Don't Make Mistakes with Tospoviruses!

Margery Daughtrey

By now you've probably learned the basics of tospovirus biology. You've heard that there are two closely related tospoviruses, impatiens necrotic spot virus (INSV) and tomato spotted wilt virus (TSWV), which are both vectored by the western flower thrips. You may have learned that tospovirus symptoms can easily fool you, mimicking everything from fungal crown rots and cankers to spray injuries. Given all of this background, the next step is to expertly plan your greenhouse sanitation program, so that you won't inadvertently be inviting virus losses. The following is a shortlist of dangerous mistakes that growers have been known to make.

Mistake No. 1: Holding on to Infected Crops

If you have found a tospovirus outbreak in your greenhouse, try to completely clean house before starting spring production. Plants such as cyclamen or holiday cactus that are grown for winter holiday sales are sometimes kept until Valentine's Day and beyond. The risk of this is clear: Not every infected plant will have symptoms that make it an obvious discard item. A faint mottling on a holiday cactus or a few round, brown spots on a cyclamen could easily escape your notice. If these holiday crops are brought into contact with your spring bedding plant crop, they may serve as a source of inoculum that could devastate your impatiens, begonia, *Browallia,* and primula, or any other susceptible crops. Crops grown from seed will escape tospovirus infection if they're kept away from thrips that have access to other infected plants. (The virus isn't transmitted through seed.) Clearing out all fall crops from greenhouses where spring crops will be grown is a good practice for thrips and tospovirus management.

Mistake No. 2: Tolerating Weeds in the Greenhouse

Weeds in the greenhouse matter much more than they ever did before. In the years before tospoviruses began to plague the greenhouse industry, educators would threaten that aphids might move viruses from weeds to crops. This was and is a theoretical possibility—but now weeds present a more specific threat. Because tospoviruses have such broad host ranges, they threaten most flower crops and they also affect many weed species. This includes many common weeds likely to be found in your greenhouse, such as oxalis and chickweed. A virus epidemic may be going on under the bench at the same time the disease is rampant in a crop. Your weeds don't even have to look sick to be a source of virus for thrips.

Mistake No. 3: Keeping Diseased Stock Plants

No one means to keep diseased stock plants, but sometimes a bed of an easy-to-propagate species has become such a fixture in the greenhouse that no one gives it thought. Whether growing pots or lounging under a bench, plants maintained in the greenhouse indefinitely may suddenly turn into a source of tospovirus. Adult thrips may infect stock with INSV or TSWV from another crop at any time, causing the previously dependable stock plants to become a major liability. One notable example of this "Typhoid Mary" stock plant concept is Swedish ivy, often kept within the greenhouse from season to season. When Swedish ivy is infected with INSV, it shows symptoms, but who would ever guess that they were virus symptoms? Dark brown lesions occur on the leaves, but the areas of dead leaf tissue aren't arranged in any pattern that causes the appropriate suspicion—if anything, the grower might suspect an injury from high salts, pesticide application, or air pollution.

Mistake No. 4: Blaming Yourself

Being good at disease diagnosis requires an observant nature and being able to notice when a plant is growing in an unusual way. Next, it's important to review the cultural conditions that any funny-looking plants have been exposed to, to see if there is a logical cultural explanation for symptoms. Then, you'll need to consider the pattern of the injury within the green-house, looking for correlations with areas that vary in moisture, air movement, proximity to the furnace, or other factors. But don't assume that there must be a logical explanation for symptoms somewhere within your cultural techniques—keep in mind that during the past decade peculiar symptoms in crops have often been due to INSV or TSWV. Many growers have been slow to obtain a positive diagnosis of a tospovirus because they didn't realize they were dealing with a contagious disease rather than a cultural problem. Because the symptoms of INSV or TSWV are so often nondescript dead, brown or black areas on leaves or stems, they're repeatedly misdiagnosed. Only by remembering to ask the question, "Is this perhaps a tospovirus symptom?" will you be able to break away from the mistake of assuming that it "must be something you've done wrong."

Pest Control, March 1998.

Index

Page numbers for photographs are **bold**, and page numbers for tables are *italic*.

Is Your "GrowerTalks on" Series in Need of Completion?

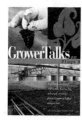